Land, Soil and Water Resources

Land, Soil and Water Resources

Mahendra Kamalvanshi

RANDOM PUBLICATIONS
NEW DELHI (INDIA)

Land, Soil and Water Resources

ISBN 978-93-5111-603-5

Published in 2015 in India by

RANDOM PUBLICATIONS

4376-A/4B, Gali Murari Lal, Ansari Road
New Delhi-110 002
Phone : +9111-43580356, 011-23289044, 011-43142548
e-mail: sales@randompublications.com,
info@randompublications.com, randomexports@gmail.com

Reprinted 2021

Type Setting by : Friends Media, Delhi-110089
Digitally Printed at : Replika Press Pvt. Ltd.

Preface

Natural resources occur naturally within environments that exist relatively undisturbed by humanity, in a natural form. A natural resource is often characterized by amounts of biodiversity and geodiversity existent in various ecosystems. Natural resources are derived from the environment. Some of them are essential for our survival while most are used for satisfying our needs. Natural resources may be further classified in different ways.

Land is among the most important natural resources. It covers only about thirty per cent of the total area of the earth's surface and all parts of this small percentage are not habitable. The uneven distribution of population in different parts of the world is mainly due to varied characteristics of land and climate. The thin layer of grainy substance covering the surface of the earth is called soil. It is closely linked to land. Landforms determine the type of soil. Soil is made up of organic matter, minerals and weathered rocks found on the earth. This happens through the process of weathering. The right mix of minerals and organic matter make the soil fertile. Water is a vital renewable natural resource. Three fourth's of the earth's surface is covered with water. It is therefore appropriately called the 'water planet'. Even today, the oceans cover two-thirds of the earth's surface and support a rich variety of plant and animal life. The ocean water is however saline and not fit for human consumption. Fresh water accounts for only about 2.7 per cent. Nearly 70 per cent of this occurs as ice sheets and glaciers in Antarctica, Greenland and mountain regions. Due to their location they are inaccessible. Only 1 per cent of freshwater is available and fit for human use. It is found as ground water, as surface water in rivers and lakes and as water vapour in the atmosphere.

I would like to thank my team for standing beside me throughout my career and writing this book. My special thanks go to "Random Publications" who have published the book.

– Mahendra Kamalvanshi

Contents

1

Biochemistry of Plants and Soils

Plants have a proclivity for perfuming the atmosphere around them. Since antiquity it has been known that both floral and vegetative parts of many species emit substances with distinctive smells. The discovery of the gaseous hormone ethylene 70 years ago brought the realization that at least some of the compounds emitted may have physiological significance without any distinctive smell to humans. At present, more than 1,000 low M_r organic compounds have been reported to be emitted from plants, although a comprehensive list is available only for floral volatiles.

The occurrence and distribution of plant volatiles has been significantly extended in the last 15 years thanks to the adoption of simple, sensitive methods for headspace sampling and the availability of relatively inexpensive bench-top instruments for gas chromatography-mass spectrometry. The substances reported are largely lipophilic products with molecular masses under 300. Most can be assigned to the following classes (in order of decreasing size): terpenoids, fatty acid derivatives including lipoxygenase pathway products, benzenoids and phenylpropanoids, C_5-branched compounds, and various nitrogen and sulfur containing compounds. Nearly all of these classes are emitted from vegetative parts as well as flowers, and some are even emitted from roots. A major discovery of the last decade is that plants commonly emit much greater amounts and varieties of volatiles after herbivore damage, and not just from the site of injury.

Plant volatile research, as in other areas of plant biology, has come from the use of molecular and biochemical techniques. A large number of genes encoding enzymes of volatile biosynthesis have recently been reported. In vitro characterization of the heterologously expressed enzymes, especially determination of their substrate and product specificity, has helped clarify the pathways of volatile formation. In addition, investigation of the spatial and temporal patterns of gene expression has provided new information on the factors regulating the emission of plant volatile compounds. In this update, we survey the latest advances on the biosynthesis and regulation of plant volatiles, beginning with a brief review of the function of these substances.

MEANING OF PLANT VOLATILES

The greatest mysteries surrounding volatiles concern their function in the life of the plant. While it is generally assumed that compounds emitted from flowers serve to attract and guide pollinators, only scattered attempts have been made to demonstrate the ability of individual substances to attract specific pollinators. Many floral volatiles have anti-microbial or anti-herbivore activity, and so could also act to protect valuable reproductive parts of plants from enemies. Among vegetative volatiles, the most intensively studied substance is isoprene, a simple five-carbon terpene emitted from the foliage of many woody species. The function of isoprene is still controversial, and this compound may act to increase the tolerance of photosynthesis to high temperatures by stabilizing the thylakoid membranes or by quenching reactive oxygen species. The release of volatiles from vegetative organs following herbivore damage seems to be a general property of plant species. Contributions to this special issue cover herbivore-induced volatiles from cabbage, cucumber, *Lotus japonicus*, and maize. These substances have been demonstrated to serve as indirect plant defences. That is, they attract arthropods that prey upon or parasitize herbivores, thus minimizing further damage to plant tissue. In some cases, herbivore-induced volatiles may also act as direct defences, repelling or intoxicating herbivores and pathogens. The possibility that these substances also act in plant-plant communication has been discussed.

Herbivore-induced volatiles could additionally have physiological roles within the plant, with their release being a consequence of their volatility and membrane solubility. Like isoprene, some herbivore-induced monoterpenes and sesquiterpenes have the potential to combine with various reactive oxygen species, and so could protect against internal oxidative damage. In fact, ozone fumigation has recently been reported to promote the emission of herbivore-induced volatiles. Yet, it is still unclear why oxidative stress is likely to be significantly higher after herbivore damage. Further studies are needed to help elucidate the roles of these and other plant volatiles. The growing number of reports on genes involved in volatile formation, as described in the following sections, should enable investigators to manipulate volatile emission and test its function in plants.

BIOSYNTHESIS OF VOLATILE TERPENES

The largest class of plant secondary metabolites, have many volatile representatives. The majority of hemiterpenes (C_5), monoterpenes (C_{10}), sesquiterpenes (C_{15}), and even some diterpenes (C_{20}) have high enough vapor pressures at normal atmospheric conditions to allow significant release into the air. The basic pathway of volatile terpenoid biosynthesis is conveniently treated in three phases: (1) formation of the basic C_5 units, (2) condensation of two or three C_5 units to form C_{10}, C_{15}, or C_{20} prenyl diphosphates, and (3) conversion of the resulting prenyl diphosphates to end products. The formation of basic C_5 units, isopentenyl diphosphate (IPP) and dimethylallyl diphosphate (DMAPP) proceeds via two alternative pathways: the long known mevalonate pathway from acetyl-CoA and the methylerythritol phosphate pathway from pyruvate and glyceraldehyde-3-phosphate, discovered only in the last 10 years. The methylerythritol phosphate pathway, localized in the plastids, is thought to provide IPP and DMAPP for hemiterpene, monoterpene, and diterpene biosynthesis, while the cytosol-localized mevalonate pathway provides C_5 units for sesquiterpene biosynthesis. However, metabolic "cross-talk" between the two pathways is prevalent, particularly in the direction from plastids to cytosol. This issue contains two contributions concerning the regulation of the basic pathways in relation to isoprene formation. Although produced largely by the plastidial pathway, isoprene also seems to arise from extra-plastidial sources, but there is apparently no cross-talk between the two pathways in its formation. The plastidial pathway is controlled by tight feedback regulation on its first step, deoxyxylulose-5-phosphate synthase.

In the subsequent phase of terpene biosynthesis, IPP and DMAPP condense to form geranyl diphosphate (GPP), farnesyl diphosphate (FPP), and geranylgeranyl diphosphate, the precursors of monoterpenes, sesquiterpenes, and diterpenes, respectively. These reactions are catalyzed by short-chain prenyltransferases. FPP is synthesized by a large family of homodimeric prenyltransferases called FPP synthases. However, the situation regarding GPP formation is more complex. While the GPP synthases of Arabidopsis and grand fir are homodimers, like other short-chain prenyltransferases, those reported from peppermint (*Mentha* × *piperita*) leaves and the flowers of snapdragon (*Antirrhinum majus*) and *Clarkia breweri* are unusual heterodimeric enzymes, with each subunit being a member of the prenyltransferase protein family.

The third phase of terpene volatile biosynthesis involves the conversion of the various prenyl diphosphates, DMAPP (C_5), GPP (C_{10}), FPP (C_{15}), and geranylgeranyl diphosphate (C_{20}), to hemiterpenes (isoprene and 2-methyl-3-buten-2-ol), monoterpenes, sesquiterpenes, and diterpenes, respectively. These reactions, carried out by a large family of enzymes known as terpene synthases, produce the primary representatives of each skeletal type. The investigation of terpene synthases is a very active area of plant volatile research and this issue contains four contributions describing the isolation of genes of this type from Norway spruce, Arabidopsis, cucumber, and *L. japonicus*. These gene sequences give new insights into the evolutionary origin and genetic regulation of terpene synthases. One of the most outstanding properties of these enzymes is their proclivity for making multiple products from a single substrate. Hence, there has been much curiosity about the carbocationic reaction mechanism. The elucidation of the first crystal structures of plant terpene synthases now puts this work on a much stronger experimental footing. Many terpene volatiles are direct products of terpene synthases, but others are formed through transformation of the initial products by oxidation, dehydrogenation, acylation, and other reaction types. These are discussed in the following section.

ALTERATION REACTIONS THAT ENHANCE THE VOLATILITY OF COMPOUNDS

The terpene pathways are essentially biosynthetic, building up a carbon skeleton, and the immediate products formed by the large family of terpene synthases discussed above are mostly hydrocarbons, although sometimes they contain a hydroxyl group (*e.g.* linalool synthase produces linalool, a tertiary alcohol). Such compounds are already fairly volatile. In contrast, most other volatile compounds are produced through the shortening of a carbon skeleton, often followed by further modification, or simply by modification of the existing carbon skeleton. Compounds that are already somewhat volatile may also be modified, resulting in enhanced volatility or changed olfactory properties. The majority of these modifications involve the reduction or removal of carboxyl groups, the addition of hydroxyl groups, and the formation of esters and ethers. Each type of modification is catalyzed by a group (or several groups) of related enzymes constituting protein families. Some of these protein families had been previously recognized from biochemical research into non-volatile compounds but some were only recently identified as part of the research into the biosynthesis of plant volatiles. Modifications for which enzymatic reactions and enzymes have been identified in plants are described below.

Corrosion by Cytochrome P450 Enzymes

The P450 cytochrome oxidases have been well characterized from a multitude of plant and animal species, and are involved in numerous metabolic pathways. Not surprisingly, these enzymes have been found to be involved in many of the reactions of volatile biosynthesis.

The basic skeleton of the monoterpenes and sesquiterpenes, discussed above, is often modified by hydroxylation. For example, 3-hydroxylation of limonene by a P450 enzyme is the first step in the biosynthesis of menthol, a volatile flavour compound found in mint, whereas a 6-hydroxylation of limonene by another P450 enzyme is the first step in the biosynthesis of another volatile spice, carvone, in the caraway (*Carum carvi*) fruit. A P450 enzyme is also responsible for the conversion of the sesquiterpene 5-epi-aristolochene to capsidiol, a dihydroxylated volatile compound. Some homoterepene compounds, for example the C11 compound 4,8-dimethyl-1,3,7-non-atriene which is often emitted from injured tissues, are believed to be derived from terpenes by cleavage catalyzed by P450 enzymes, but such enzymes have not yet been identified conclusively.

Cytochrome P450 enzymes are also very important in the biosynethsis of volatile phenylpropenes such as eugenol and the benzenoid vanillin. Both of these compounds, found in a wide variety of species both in flowers and in leaves, are derived from Phe and share with non-volatile phenylpropanoids the earlier steps of 4-hydroxylation of cinnamate by 4CH, a P450 enzyme, and 3-hydroxylation by the newly discovered P450 enzyme that utilizes the shikimic or quinic ester of coumarate rather than coumaric acid or coumaroyl-CoA.

Cytochrome P450 enzymes are crucial in the biosynthesis of volatiles derived from fatty acids, and in particular, in the octodecanoic pathway. Two different P450 enzymes, 9-LOX and 13-LOX, can introduce a peroxide into linoleic acid at the respective positions. Subsequent cleavage of the hydrocarbon chain by hydroperoxide lyases produces non-adienal and 3-cis-hexenal, respectively. The later is an important component of green leaf volatiles, the mixtures of compounds that are emitted when the leaf is damaged. In addition to the C6 aldehyde, cleavage of linoleic acid at the 12 to 13 bond also produces a C12 compound that can subsequently be converted to jasmonic acid. While jasmonic acid is not by itself volatile, its methylester.

Corrosion by Dehydrogenases

NADP/NAD-dependent oxidoreductases are another large and well-studied family of proteins with representatives found to be involved in the biosynthesis of volatiles. Such enzymes have been implicated in the interconversion of volatile alcohols and aldehydes. For example, apparently non-specific alcohol dehydrogenases can convert short-chain aldehydes such as hexanal and 3-cis-hexenal to hexenol and 3-cis-hexenol, alcohols that are also found in damaged leaves. This lack of tight substrate specificity was used to alter the aroma profile of ripe tomato (*Lycopersicon esculentum*) fruit by genetic engineering. Some terpene alcohols such as geraniol and carveol are converted to aldehydes by similarly non-specific dehydrogenases. Geranial and neral (which are coproduced by the oxidation of geraniol; the mixture is termed citral), have lemony aroma and are found in many plants, and carvone gives caraway its distinct flavour. And benzyl alcohol, a major floral scent component in many Nicotianeae species and elsewhere, is also likely derived from benzaldehyde in a reversible reaction catalyzed by a member of the NADP/NAD-dependent oxidoreductases family.

Methylation of Hydroxyl Groups

A large portion of plant volatiles contain a methylated hydroxyl group (*i.e.* a methoxyl group). The methyl group is usually added in a reaction catalyzed by a methyltransferase (MT) in which *S*-adenosyl-L-methionnine (SAM) serves as the methyl donor. All plant methyltransferases appear to share a similar SAM-binding domain; however, they fall into distinct families that share little primary sequence similarity elsewhere. A large family of methyltransferases with members involved in the synthesis of both volatile and non-volatile "small molecules" (as opposed to proteins or nucleic acids) has been identified and designated as the Type I methyltransferase family. Members of this MT family have been shown to catalyze the 4-hydroxyl methylation of eugenol to form methyleugenol in flowers of *C. breweri* and in the glands of basil (*Ocimum basilicum*), and also the methylation of chavicol to methylchavicol in the basil glands. 3,5-Dimethoxytoluene, a major scent compound in many hybrid roses, is produced from orcinol (3,5-dihydroxytoluene) in two successive methylation reactions catalyzed by two very similar MTs, orcinol OMTs. Both enzymes can carry out both reactions; however, OOMT1 is more catalytically efficient with orcinol while OOMT2 is more catalytically efficient with 3-methoxy,5-hydroxytoluene. Chinese rose (*Rosa chinensis*) flowers make a similar compound with three methoxyl groups, 1,3,5-trimethoxybenzene, which is synthesized from 1,3,5-trihydroxybenzene. OOMT1 and OOMT2 can catalyze the methylation of the second and third intermediates (1-methoxy,3,5-dihydroxybenzene and 1,3-dimethoxy,5-hydroxybenzene) but not the methylation of 1,3,5-trihydroxybenzene, also known as phloroglucinol. The enzyme that methylates this compound, phloroglucinol OMT (POMT), also belongs to the Type I MT family, but is only distantly related to OOMT1 and OOMT2.

An important strawberry (*Fragaria* × *ananassa*) aroma compound, 2,5-dimethyl-4-methoxy-3(2H)-furanone, was shown to be produced by the action of another Type I MT. This MT appears to be able to methylate a wide range of substrates, including intermediates of the lignin biosynthetic pathway such as coniferal aldehyde and coniferal alcohol. Eugenol, mentioned above as the substrate of a eugenol MT and an important volatile spice on its own, and vanillin, another important aroma compound, also contain a 3-methoxyl group on their benzene ring. The synthetic pathways of these compounds share the first few steps with the lignin pathway. Gang *et. al.* demonstrated that eugenol is derived from a lignin intermediate past the para and meta hydroxylation reactions and the 3-hydroxyl methylation, which is catalyzed by CCOMT (caffeoyl-CoA OMT). This is also likely to be the case for vanillin, although this has not yet been demonstrated conclusively. CCOMT is a member of the Type II MT family of plants.

Methylation of Carboxyl Groups

Some methyl esters are extremely wide spread in the plant kingdom. For example, methylsalicylate has been reported in numerous floral scents, and it is also commonly emitted from vegetative tissues under attack by insects or parasites. An enzyme capable of methylating salicylic acid (SA), salicylic acid carboxyl methyltransferase (SAMT) was first reported from *C. breweri* flowers.

It has since been identified from several other plant species. This enzyme, which uses SAM as the methyl donor, defines a new type of plant MT known as Type III or SABATH MT (after the first two letters of the names of the first three enzymes identified in this family). Some SAMT enzymes have been shown to be able to methylate also benzoic acid (BA), a compound identical to SA except for lacking the 2-hydroxyl group present in SA. On the other hand, benzoic acid carboxyl methyltransferase (BAMT), the enzyme responsible for the snapdragon floral volatile methylbenzoate, cannot methylate SA.

There are 24 SABATH MTs encoded by the Arabidopsis genome, including one enzyme that methylates both SA and BA. It is not known if all of these MTs are involved in the biosynthesis of volatile compounds, but MTs belonging to the SABATH family have been shown to be responsible for the three consecutive metylation reactions in the biosynthesis of caffeine, a non-volatile compound, in *Coffea arabica*. However, another Arabidopsis SABATH MT was shown to methylate jasmonic acid to form methyljasmonate. While this molecule may act as an internal signal molecule in Arabidopsis and other plant species, it is also emitted from injured plants, and has also been reported in the floral scent of several plant species where it is likely to be formed by similar enzymes.

Acylation

Acylation, most often with an acetyl moiety but also with larger acyls such butanoyl or benzoyl acyls, to make volatile compounds is also common. In all known examples, such plant volatile esters are synthesized by a recently discovered family of plant acyltransferases called BAHD, after the first letter of the first four enzymes identified. The basic reaction catalyzed by these enzymes is the transfer of an acyl group from an acyl-CoA intermediate to the hydroxyl group of an alcohol. Many BAHD enzymes are involved in the synthesis of non-volatile compounds such as acylated alkaloids or taxol derivatives, or in early steps of pathways that may lead to the synthesis of volatiles such as eugenol. BAHD acyltransferases directly involved in volatile synthesis include benzyl alcohol acetyl-CoA transferase from *C. breweri* flowers, which produced benzyl acetate; benzyl alcohol benzoyl-CoA transferase, which produces benzylbenzoate in flowers of Clarkia and petunia, and in tobacco mosaic virus-infected leaves of tobacco; and 3-cis-hexen-1-ol acetyl CoA transferase, which produces 3-cis-hexenyl acetate (a green leaf volatile) and is induced in damaged leaves of Arabidopsis.

The BAHD enzymes often show wide substrate specificity for both the acyl moiety and the alcohol moiety. For example, the petunia benzyl alcohol benzoyl-CoA transferase enzyme can also transfer an acetyl moiety to the alcohol phenylethanol, producing phenylethylacetate. Similarly, a BAHD enzyme from ripening strawberry (*Fragaria* spp) fruit, can use a series of acyl moieties such as acetyl, butanyl, and hexanyl, and transfer them to various alcohols such as heptanol, octanol, and geraniol. An acyltransferase from banana (*Musa sapientum*) has similarly wide substrate specificity, and a rose (*Rose hybrida*) flower BAHD enzyme can acetylate both geraniol and citronellol. In such cases, the type of volatile formed in a given tissue depends more on the internal concentrations of the substrates than on the K_m and K_{cat} values of the enzyme for these substrates.

The Manufacture of the C6-C1 Benzenoids from C6-C3 Phenylpropanoids

The shortening by two carbons of the three-carbon chain attached to the phenyl ring of phenylpropanoids leads to the formation of benzenoid compounds. The mechanism by which this is achieved is not fully understood. In vivo stable isotope labeling and computer-assisted metabolic flux analysis, described in this issue, revealed that both the CoA-dependent—oxidative and CoA-independent-non—oxidative pathways are involved in the formation of benzenoid compounds in petunia. However, a recent discovery also indicates that in the case of 2-hydroxybenzoic acid (*i.e.* salicylic acid), a third pathway, via the isochorismate pathway, may also operate in plants, as it does in bacteria

GUIDELINE OF EMISSION OF VOLATILE COMPOUNDS

Emission of a particular volatile compound into the atmosphere depends on both the rate of its biosynthesis and the rate of its release. Substantial progress in the last decade in the isolation and characterization of genes responsible for the formation of volatile compounds has facilitated the investigation of the regulation of the biosynthesis of plant volatiles. It has been found that volatiles are synthesized de novo in the tissues from which they are emitted. Biosynthesis normally occurs in the epidermal cells of plant tissues from which they can escape into the atmosphere or rhizosphere after being synthesized or in the secretory structures or glandular trichomes as, for instance, was found in peppermint, *Artemisia annua*, and sweet basil.

Formation of volatile compounds is spatially regulated. Of the plant organs in scented species, flowers produce the most diverse and the highest amount of volatile compounds, which peak when the flowers are ready for pollination. Vegetative tissue also releases small quantities of volatile organic compounds, which could be induced by mechanical damage or by herbivore- or pathogen infection. In herbs, such as peppermint, significant amounts of volatile compounds accumulate in the leaf glandular trichomes and the emitted volatiles represent only a small fraction of the total pool produced.

Production and emission of volatile compounds is also a developmentally regulated process. Volatile emission in flowers and accumulation in leaves and fruits follow similar developmental patterns, increasing during the early stages of organ development (when leaves are young and not fully expended, fruit is not yet mature, or when flowers are ready for pollination) and then either remaining relatively constant or decreasing over the organs' lifespan. The concurrent temporal changes in activities of enzymes responsible for the final steps of volatile formation, enzyme protein levels, and the expression of corresponding structural genes suggest that the developmental biosynthesis of volatiles is regulated largely at the level of gene expression. It is still unclear to what extent transcriptional, posttranscriptional, translational, posttranslational, and other events contribute to this process.

In general, more than one biochemical pathway is responsible for a blend of volatile compounds released from different plant tissues. A comparative analysis of the regulation of benzenoid and monoterpene emission in snapdragon flowers revealed that the orchestrated emission of phenylpropanoid and isoprenoid compounds is regulated upstream of individual metabolic pathways and includes the coordinated expression of genes that encode enzymes involved in the final steps of scent biosynthesis. However, transcription factors that regulate multiple biosynthetic pathways leading to the formation of odour bouquet have not yet been discovered.

The level of the enzyme responsible for the final step of the biosynthesis of a particular volatile is not the only limiting factor. The target for the regulation of developmental production of volatile compounds also includes the level of supplied substrate in the cell. In the case of enzymes that are able to use several similar substrates, such as SAMT and acyltransferases, the level of supplied precursor controls the type of produced product. The role of substrate in the regulation of the biosynthesis of volatile compounds was also recently confirmed by metabolic engineering. When the linalool synthase gene was introduced under the control of the cauliflower mosaic virus 35S constitutive promoter into petunia W115, the differences between organs in the amount of the synthesized linalool or its glycoside depended more on the availability of the substrate GPP in the tissue than on the expression of the linalool synthase gene. In peppermint the up-regulation of 1-deoxy-D-xylulose-5-phosphate reductase, which catalyzes the conversion of 1-deoxy-D-xylulose-5-phosphate to methylerythritol phosphate, increased the flux to GPP and led to about a 50 per cent increase in the essential oil production. Additional regulation of GPP formation can occur at the level of GPP synthase, as was shown in snapdragon where the small subunit can play a key role in GPP biosynthesis. Feedback regulation of GPP synthase by product and substrate inhibition could also contribute to the regulatory control of the flux to GPP and subsequently to monoterpene production. When enzymes competed for the same substrate as in the case of three monoterpene synthases introduced into tobacco plants, the magnitude of monoterpene emission in leaves was close to that predicted based on the K_m values of the enzymes for GPP, while the emission levels in flowers were comparable suggesting that the GPP pool did not limit monoterpene production. These results show that while the investigation of the regulation of the final steps of volatile biosynthesis is an important starting point, a detailed understanding of the regulation of the flux through the entire biochemical pathway is essential for the complete understanding of production and emission of secondary volatile compounds.

Emission of volatile compounds from flowers and leaves of some plant species, as well as herbivore-induced volatiles, varies remarkably throughout the photoperiod. The release of floral volatiles in these species displays a rhythmic pattern with maximum emission during the day or night, which generally coincides with the foraging activities of potential pollinators, and is controlled by a circadian clock or regulated by light. Isoprene, as well as volatiles emitted from undamaged and herbivore-attacked leaves, exhibit a distinct diurnal emission pattern with some leaf volatiles' emission also controlled by a circadian clock—pinene in*Artemisia annua*. In flowers the rhythmic production and emission of some volatile compounds, such as the volatile ester methylbenzoate, is regulated primarily by the level of substrate availability (BA), which in turn could be regulated at the level of expression of genes encoding the key enzymes of its biosynthesis. While regulation of isoprene emission is well understood, little is known to date about the molecular mechanisms responsible for diurnal emission of inducible vegetative volatiles. Often, there is a delay of several hours between the beginning of herbivore damage and the release of induced volatile compounds, which could be explained, in part, by the up-regulation of the expression of genes responsible for their biosynthesis. When volatile compounds are released immediately after herbivore damage, they arise from stored pools.

Environmental factors such as light, temperature, and moisture status can greatly influence the emission of volatiles and the yield and composition of essential oils. In addition, pollination induces the decrease in emission of floral volatiles which begins after pollen tubes reach the ovary. To date very little is known about the release of synthesized volatile compounds from plant tissues. In general, the rate of release is a function of the physical properties of the compound itself (its volatility) and the properties of cellular and intracellular membranes (in case of monoterpenes which are synthesized in the chloroplast) through which the compound has to diffuse. Comparative analysis of volatile compounds emitted and present within the plant tissue revealed that the emission of volatiles is not merely a function of their differential volatility but could also involve a cytologically organized excretory process. The membranes of the storage compartment (where it exists) or epidermal cell wall might be selectively more permeable to some volatile compounds or the emitted substances may be associated with an entirely different secretory compartment than the stored volatiles. Virtually nothing is presently known about metabolite trafficking between various subcellular compartments, the mechanism of the release process, and how these processes contribute to the regulation of volatile emission.

SOIL BIOCHEMISTRY

The soil biochemist is concerned with the integration of two vast areas of study: soils and biochemistry. Physically a soil is a porous aggregate of mineral particles containing a small amount of organic material and living organisms. Contained within the soil pores are the soil gases and water. A silt loam surface soil has approximately 50 per cent of its total volume as solids and 50 per cent as pore space. Only 5 per cent of the total soil volume (10 per cent of the solid volume) is made up of carbon-derived material that is the element upon which our biochemical system is built. However, the importance of this fraction is much greater than its size would first indicate.

The majority of soil biochemical reactions deal primarily with the organic fraction. The organic fraction of soil serves as a storehouse of essential nutrients for plant and microbial growth, is a major source of energy for microorganisms, and plays an important role in determining soil structure.

Soil organic matter consists of (i) newly deposited organic materials, (ii) partially decomposed organic materials, (iii) the more chemically resistant humus, and (iv) the microbial biomass. The biochemical reactions associated with the organic matter fraction are greatly affected by shifts in environmental factors.

The pore space occupied by air and water will vary, but at optimum moisture for plant growth 40 per cent of the pore space contains water. As water interacts with the solid portion of the soil, it brings about various reactions resulting in the "soil solution". It is from the soil solution that plants and microorganisms obtain their essential nutrients. The soil solution cannot be specifically defined because it is in a constant state of flux and large variations in the composition of water located in different places in soil occurs. Generally, it can be considered a rather weak salt solution on the order of approximately 0.01 molar in concentration.

CHEMICAL PROBLEMS OF SOILS

All landuse activities, particularly those which are poorly managed, involve destruction or disturbance, to a greater or lesser extent, of natural and semi-natural ecosystems. Almost invariably, however, it is these ecosystems, in equilibrium with their environment, which offer most effective protection to the soil which supports them. A major consequence of ecosystem destruction and disturbance is that of soil degradation.

This has been defined as the decline in soil quality caused through its misuse by human activity. More specifically, it refers to the decline in soil productivity through adverse changes in nutrient status, organic matter, structural stability and concentrations of electrolytes and toxic chemicals.

Soil degradation incorporates a number of environmental problems, some of which are interrelated, including erosion, compaction, water excess and deficit, acidification, salinisation and sodification, and toxic accumulation of agricultural chemicals and urban/industrial pollutants. In many instances, these have led to a serious decline in soil quality and productivity, and it is only in recent decades that the finite nature of soil as a resource has become widely recognised.

Soil degradation is not a new phenomenon. Archaeological evidence suggests that it has been on-going since the beginning of settled agriculture several thousand years ago. The decline of many ancient civilizations, including the Mesopotamians of the Tigris and Euphrates valleys in Iraq, the Harappans of the Indus valley in Pakistan and the Mayans of Central America, was due in part to soil degradation. More recently, an event of major significance was the dustbowl which occurred in the Great Plains of the American midwest during the 1930s.

At this time, intensive agricultural practices, employed in the eastern states, were transferred to the drier midwest where the soils are lighter textured and more susceptible to erosion. A number of years of drought, combined with crop failure and destruction of the protective organic-rich topsoil, resulted in severe wind erosion.

The effects of soil degradation are not restricted to the soil alone, but have a number of off-site implications. Soil erosion, for example, is often associated with increased incidence of flooding, siltation of rivers, lakes and reservoirs, and deposition of material in low-lying areas. These problems may be compounded in areas where infiltration capacity is reduced due to compaction, hardsetting or induration of soils.

Salinisation and sodification of soils are often associated with poor quality irrigation water, while soil acidification is commonly linked with acidification and aluminium contamination of surface waters. Leaching of fertilizers and pesticides from agricultural soils may also lead to contamination of surface and shallow ground waters. In addition, contamination of soils by urban and industrial pollutants, such as heavy metals and radionuclides, may lead to toxic accumulation in arable produce and in herbage for grazing animals, thus having important implications for human health.

The extent of soil degradation is influenced by a number of factors, many of which are interrelated, namely soil characteristics, relief, climate, land use, and socio-economic and political controls. In many studies of soil degradation and its wider environmental implications, the socio-economic and political controls are often overlooked, or at least not examined in any detail, perhaps because of the difficulties associated with the collection of reliable and comparable data. Increasingly, however, these controls on land use systems are being viewed as central to the issue of soil degradation, particularly in the developing world.

Management of soil degradation, whether at a global, regional or local scale, is clearly a complex issue and represents one of our most challenging environmental problems. Emphasis should be placed on sustainable rather than exploitative landuse practices; this theme was highlighted by the World Soil Charter which called for a commitment by governments, agencies and land users to 'manage the land for long term advantage rather than short term expediency'. The problem requires a holistic, multidisciplinary approach involving the collaborative and co-ordinated efforts of ecologists, agronomists, soil scientists, hydrologists, engineers, sociologists and economists.

Moreover, the involvement of government and non-government organisations, aid agencies and the farmers themselves is essential to the success of research and development in this area. Such involvement should facilitate the implementation of education, training and incentive programmes.

Imposition from above of high-technology, high-cost solutions by technical experts from developed countries is certainly not the answer in the developing world. Inevitably such solutions are not economically viable and low-technology, low-cost options, such as low external input agriculture, agroforestry and social forestry, are often the only answer.

Hence, the approach to soil conservation has shifted in recent years from a rather technocentric standpoint to a more ecocentric position. Central to this approach are the concepts of land husbandry and sustainable development, which place emphasis on the land users themselves rather than on the technical experts and advisors.

The most pressing soil degradation problems, and in each case the causal factors, on- and off-site effects, and management strategies will be considered.

PHYSICAL PROBLEMS

Erosion

Soil erosion occurs when the rate of removal of soil by water and/or wind exceeds the rate of soil formation. Generally, rates of soil formation are very low, with profiles developing at a rate of about 1 cm every 100-400 years; assuming an average bulk density of 1.33 g/cm^3, this equates to about 0.3-1.3 t h. It is important to differentiate between natural or background erosion, and erosion which has been accelerated largely as a result of human activity.

Background erosion rates are often similar to rates of soil formation at <1.0 t h, although in mountainous areas they may be considerably higher. In contrast, rates of accelerated erosion commonly exceed 10 t h and sometimes exceed 100 t h. Some of the highest soil erosion rates have been observed in the loess plateau area of China and in the Himalayan foothills of Nepal, where values in excess of 200 t h have been recorded. Similarly, in India, gully erosion results in a loss of about 8,000 ha of land per year.

The extent of soil erosion is governed by a number of factors. Those of particular importance include erosivity of the eroding agent, erodibility of the soil, slope steepness and length, landuse practices and conservation strategies. These factors are summarised in the Universal Soil Loss Equation which has been used widely in the modelling and prediction of soil erosion inter-rill erosion and is not easily applied to areas where gully and stream bank erosion are widespread.

Its universal nature has also been questioned, particularly in terms of its application to tropical soils. Furthermore, it should be emphasised that this model does not consider the wide range of socio-economic and political factors which play a crucial role in terms of their influence on the degree of soil erosion; these will be examined later.

Alternative models include SLEMSA (soil loss estimator for southern Africa) and CREAMS (chemicals run-off and erosion arising from agricultural management systems).

Erosivity is a measure of the potential of the eroding agent to erode and is commonly expressed in terms of kinetic energy. The erosivity of rainfall relates to the detaching power of raindrops, which increases with increasing rainfall intensity and is largely a function of drop size. This relationship is non-linear, however, and at intensities in excess of 50 mm hr the increase in kinetic energy is minimal as turbulence prevents further increase in drop size.

Rainfall erosivity tends to be greatest in areas where there is a marked seasonal concentration of rainfall. There are a number of rainfall erosivity indices, although one of the most widely used is the EI30 index which is a compound index of kinetic energy and the maximum 30-minute rainfall intensity. Indices of wind erosivity are based largely on the velocity and duration of the wind.

Erodibility of the soil is a measure of its resistance to detachment and transport, and depends on a number of soil characteristics, particularly texture, organic content, structure and permeability. In terms of both water and wind erosion, the most erodible soils tend to be characterised by low clay and organic contents, and poor structural stability.

Similarly, soils with an intermediate texture (fine sand to coarse loam) are more erodible than both coarse-textured sandy soils, where particle size and mass are large, and fine-textured clayey soils, which are more cohesive. The erodibility factor (K) in the Universal Soil Loss Equation can be determined from soil texture, organic content, structure and permeability using the monograph.

Land use is perhaps the most significant factor influencing soil erosion, for two main reasons. First, many landuse practices leave the soil devoid of a protective vegetation cover, or with only a partial cover, for significant periods of time and second, they involve mechanical disturbance of the soil.

Specific aspects of land use often associated with accelerated soil erosion include expansion and intensification of arable cultivation, overgrazing, deforestation, certain forestry practices, site clearance in preparation for urban and industrial construction, and a number of recreational activities such as walking and skiing.

Arable cultivation has expanded and intensified dramatically in recent decades. Relatively steep slopes, formerly covered by grass or trees, have been converted to arable cropping, while an increased use of heavy agricultural machinery has resulted in compaction of the soil. This in turn has led to reduced infiltration capacity, particularly along wheel tracks, thus resulting in increased surface run-off and erosion. Similarly, increased reliance on tillage activities, throughout the cropping cycle, has rendered soils more susceptible to erosion.

This problem has been compounded by the decline in levels of soil organic matter, and hence structural stability, largely in response to increased use of inorganic fertilizers. In addition, the tendency to increase field sizes on arable land has meant that there are fewer physical breaks and barriers in the landscape, such as tree lines, hedgerows and walls, to restrict erosion.

Susceptibility to erosion is further increased if land is cultivated with the slope rather than parallel to the contours. These problems are well illustrated by Boardman in studies of soil erosion in southeast England.

Overgrazing is particularly common in drought-affected parts of the developing world, such as the Sahelian region of sub-Saharan Africa, and the rangelands and communal lands of eastern and southern Africa.

In a study of the impact of grazing on soils of the savanna region of Nigeria, attribute enhanced surface run-off and erosion to compaction of the soil and destruction of the protective vegetation cover by grazing animals, and to the adoption of inappropriate burning strategies.

Deforestation, largely for logging and woodfuel purposes, is also common in many parts of the developing world.

Trees are well known for their ability to protect soils from erosion, particularly on steeply sloping terrain. Their root systems, and the organic material which they supply, help to stabilise the soil, while water uptake and canopy interception serve to reduce the frequency and intensity of surface run-off.

In addition to deforestation, many forestry practices are associated with accelerated soil erosion, including the needleleaf forestry programmes which have become widespread in many areas of upland Britain. Here, erosion is most serious during the pre-planting stages of land preparation and drainage, and after harvesting. In relation to urban and industrial land use, construction and associated disturbance of land may lead to increased soil erosion. Even certain recreational activities have been implicated in this problem, including walking and skidding.

A number of socio-economic and political factors have been associated with accelerated soil erosion, particularly in the developing world. These include population pressure, skewed land resource distribution, poverty and marginalisation, increasing demand for wood fuel, inappropriate land tenure and farm policies, small size of land-holdings and poor infrastructure.

In many developing countries, population growth is rapid and the demand for agricultural land and wood fuel is ever increasing. Furthermore, agricultural systems are characterised by a skewed land resource distribution where a minority of affluent and powerful landowners control a majority of the land area. The poorest farmers are thus forced onto marginal land, which is particularly susceptible to erosion, and often end up in a vicious spiral of debt. Rural-urban migration, abandonment of land and increased soil erosion are often responses to this poverty trap situation.

In many parts of the developing world, large areas of land are utilised for mono-cultivation of cash crops, which are not necessarily best suited to soil conditions, rather than for indigenous mixed food cropping. Such commercial pressure on agricultural systems, as well as contributing to the problem of marginalisation discussed above, has a detrimental effect on soil quality and is unlikely to be sustainable in the long term.

There is also little political support in terms of education, training and incentive schemes to encourage farmers to adopt more sustainable landuse practices. The establishment of appropriate and comprehensive soil conservation and land husbandry programmes is further hindered by the small size of land-holdings and the large numbers of farmers involved.

The on- and off-site effects of soil erosion are considerable. At the global scale, it is estimated that unless soil conservation measures are introduced on all cultivated land, 544 million ha of potentially productive rain-fed crop land will be lost, and agricultural production will decrease by almost 20 per cent, by the year 2000.

Undoubtedly, these effects will be felt most severely in those developing countries which are least able to cope with the problem. It should be noted that the deterioration in soil productivity is disproportionate to the amount of soil eroded, as it is the nutrient-rich and structure-supporting constituents in the topsoil which are lost most readily.

The off-site impacts of soil erosion are just as serious as the on-site effects, and are often more dramatic. Numerous incidents of disruption to communication systems, flooding, siltation of water supplies and damage to property have been reported. In financial terms, the Conservation Foundation of the USA estimates the annual on-site and off-site costs of soil erosion, on a national basis, to be about $40 million and $3 billion respectively.

Allocation of responsibility and accountability for off-site incidents, and the implications for government policy and legislation, are currently the subject of debate in Britain; this situation is in marked contrast to that in many other countries, such as Canada and New Zealand, where government policies on soil conservation and farmer education are well established and rclatively successful.

In terms of the management and remediation of soil erosion, Morgan identifies three main strategies—agronomic, soil management and mechanical. Agronomic practices aim to protect the soil through sensible cropping programmes and are based on the encouragement of a dense vegetation cover and plant root network. Under these conditions, the time period over which the soil is left bare, and thus susceptible to erosion, is minimised.

An example of an agronomic practice is strip-cropping, in which alternate rows of different crops are arranged, usually across the slope, so that they grow and are harvested at different times, thus ensuring at least a partial cover of vegetation for much of the year. Using this approach, arable crops may be combined with strips of grass or trees. Retention of crop stubble in the soil, rather than burning, and the planting of trees or shrubs around the heads of gullies to reduce the rate of head ward recession, are other examples of agronomic practices.

Soil management techniques aim to increase the resistance of soil to erosion, and focus mainly on the improvement and maintenance of soil structure. Such practices include mulching, where organic residues are added to the soil, and reduced or zero-tillage options, including direct drilling, where mechanical disturbance of the soil is minimised. Soil erodibility may also be reduced by using synthetic soil conditioners.

These consist largely of organic polymers such as PVA (polyvinyl alcohol), PAM (polyacrylamide) and PEG (polyethy-leneglycol), and are applied as a surface film or are incorporated into the topsoil. Treatment is costly, however, and tends to be restricted to specialised applications such as the stabilisation of sand dune systems, and protection of the fine tilth of seedbeds.

Bio-engineering techniques, including the use of biodegradable geotextiles, have also been used to reduce the susceptibility of soils to erosion. At the Cairngorm ski slopes in Scotland, for example, these have been used successfully to facilitate the recovery of sensitive alpine plant communities.

Mechanical techniques aim to reduce the energy of the eroding agent and often involve the modification of surface topography. Examples include terracing, contour bunding and the construction of diversionary spillways to direct water away from areas which are particularly susceptible to erosion.

Shelter belts may also be planted to reduce wind erosion, although these will compete with crops for soil moisture, which may be problematic in drier areas. At the more local scale, check dams may be constructed in gully systems in an attempt to reduce erosion risk, while gabions (wire baskets filled with stones) may be installed to reduce bank erosion and headward recession. In addition to their use on agricultural land, mechanical techniques are also used to control footpath erosion.

These include the installation of drainage channels to direct water flow away from damaged stretches of footpath, and the construction of elevated sections to avoid damage to poorly drained soils. In extreme circumstances, footpaths are temporarily or permanently re-routed to facilitate the recovery of damaged stretches. In terms of the relative effectiveness of the above measures, agronomic techniques tend to offer greatest protection to the soil, as they are able to control both the detachment and transport phases of soil erosion. Soil management and mechanical techniques tend to be less effective, particularly in their control of the detachment phase of erosion.

For this reason, it is usual to combine soil management and mechanical techniques with agronomic measures. Agronomic practices are often used in preference to others, not only because they are effective in the control of erosion, but also because they are relatively cheap to adopt, can be adapted to fit in with traditional cultivation methods and are acceptable to local people.

In recent decades, there has been a change in focus away from a largely technocentric approach to soil conservation, towards a broader and more ecocentric view of land management. This places emphasis not only on the soils themselves, but also on the landuse system, the socio-economic and political controls, and the wider off-site implications of soil erosion. More important, however, is that the ecocentric approach focuses on the land users themselves and on their role in the determination of their own future.

This is in marked contrast to the imposition of decisions from above by technical experts, which perhaps explains, at least in part, the failure of many early soil conservation projects. Specific practices which have been effective in the control of soil erosion, particularly though not exclusively in the developing world, include agroforestry, social forestry and low external input agriculture.

Compaction

Soil compaction involves the compression of a mass of soil into a smaller volume and is usually expressed in terms of dry bulk density, porosity and resistance to penetration. The bulk density of soils which have been compacted by agricultural machinery, for example, may exceed 1.5 g/cm^3, while that of comparable uncompacted soils is usually between 1.0 and 1.5 g/cm^3.

The ease with which soils are compacted depends on a number of characteristics, particularly texture, occurring most readily in soils which contain appreciable quantities of clay. The compacting force, which usually acts in a vertical direction, causes alignment of the clay platelets in a direction which is more or less parallel to the ground surface. Alignment of the clay particles in this way often leads to the formation of compaction or cultivation pans which may be a few centimetres in thickness.

Such pans tend to form at a depth of about 20-30 cm, are often characterised by a well developed platy structure, and are commonly associated with impeded drainage and restricted development of plant root networks. In silty soils, raindrop impact may lead to surface crusting which is another form of soil compaction. Surface crusts may be several millimetres in thickness, and are commonly associated with reduced infiltration, increased surface run-off and accelerated soil erosion, and restricted germination and emergence of crops. In addition to textural controls, susceptibility to compaction also increases with increasing organic matter and water contents.

Closely related to soil compaction is the process of hardsetting which involves an increase in bulk density but without the application of an external load. Hardsetting soils are characterised by their low structural stability. During and after wetting, slaking and collapse of aggregates lead to uniaxial shrinkage and dispersal of clay and silt and, on drying, the soils harden without restructuring. Acceptance of hardsetting soils as a distinctive group has been hindered by the difficulties experienced in distinguishing hardsetting characteristics from other forms of soil behaviour.

One of the main causes of soil compaction is the excessive use of heavy agricultural machinery, and compaction is particularly severe along wheelings left by these vehicles. For example, bulk density values of 2.2 g/cm^3 and 1.3 g/cm^3 have been recorded in soils of wheeled and inter-wheel areas respectively. The intensification of arable cultivation in recent decades has been facilitated to a large extent by increased mechanisation.

Most procedures in the cropping cycle, from tillage and seedbed preparation, through drilling, weeding and agrochemical applications to harvesting, are now largely mechanised, particularly in the developed world. The timing of cultivation in relation to precipitation and soil water levels is known to be critical with respect to soil compaction. If soils are cultivated when they are near or above field capacity, or their plastic limit, then severe structural damage and compaction are likely to occur.

Soil compaction is not restricted to arable land but is also common on land used for grazing and forestry. Grazing animals are well known for their ability to cause compaction, or poaching, especially during wet conditions.

It is particularly common to find poorly drained or puddled areas in the vicinity of food troughs or gateways where animals tend to congregate. In forestry plantations in upland Britain, heavy agricultural machinery is often used during the preparation and drainage of land prior to planting, and during harvesting operations; such mechanisation is often associated with the disturbance and compaction of soils. Similarly, soil compaction has been reported in association with the clearance of tropical rainforest, particularly where mechanical methods are used in preference to the traditional slash and burn approach.

Soil restoration following mineral extraction, quarrying and opencast coal mining has also been implicated in the soil compaction problem. Compacted horizons result largely from the passage of heavy machinery during soil stripping and replacement, and from the shear forces produced during the lifting process.

Compaction has adverse effects on a number of soil characteristics. Increased bulk density and the consequent decrease in porosity are associated with both increased water logging and poor aeration; these changes in turn have a detrimental effect on the thermal characteristics of soils.

Under these conditions, plant growth may be restricted, particularly during the early stages of germination and emergence, and during the main phase of root network development. Similarly, the increased incidence of root rot in soils with compacted cultivation pans has been widely documented. Other characteristics that may be adversely affected by compaction include the size and diversity of soil organism populations, and the incidence of certain crop pests and diseases may also increase.

The financial impact of soil compaction is difficult to assess due to the influence of a number of interrelated factors. In addition to yield reductions, damaged soil structure may have to be restored, surface run-off and soil erosion may increase, fertilizer usage becomes less efficient due to increased leaching losses, water loss by evapouration may increase, together with the operational costs of irrigation, and in soils which are susceptible to water logging, expensive drainage systems may have to be installed. It is estimated that the cost of fuel required for tillage of compacted soils may be up to 35 per cent greater than that for uncompacted soils.

Successful management and remediation of soil compaction centres on two main approaches—improvement and maintenance of soil structure, and appropriate tillage practices. Good soil structure provides the basis for increased mechanical strength and stability, improved water retention and aeration, and protection of plant nutrients from leaching. Soil structure can be improved and maintained through the promotion of sensible crop rotation practices, particularly those which include significant periods of ley pasture, mulching and stubble retention.

Unlike continuous cropping, such practices encourage the accumulation of soil organic matter and the development of strong plant root networks, both of which are essential ingredients in the development of good soil structure. Soil structure also benefits from the addition of certain nutrients, particularly the multivalent base cations which play a crucial role in the aggregation process.

In addition to soil structural improvements, good tillage practices are an essential component of effective management and remediation of soil compaction. Tillage of agricultural land is practised for a number of reasons—first, to create a fine tilth which facilitates effective germination and emergence of seedlings, second, to remove weeds which compete with crops for nutrients, water and light, third, to incorporate crop residues into the soil with the aim of improving soil structure and nutrient content, and fourth, to improve drainage and aeration in the root zone (rhizosphere).

Although tillage is therefore of benefit to the soil, when mechanised it can lead to soil compaction. The timing of tillage is particularly important in this respect, and it should be avoided in wet conditions, particularly if soil moisture content is likely to exceed the field capacity or plastic limit. In an attempt to alleviate compaction resulting from the excessive use of heavy agricultural machinery, reduced or zero-tillage practices have been widely adopted. Such practices are inappropriate for poorly drained soils, however, where conventional tillage helps to improve drainage, aeration and the thermal characteristics of the topsoil.

They should also be avoided on excessively drained soils, where increased nitrate leaching and denitrification may lead to nitrogen deficiency, and on steep slopes where seed slots may be subjected to increased erosion.

Thus, it appears that the benefits of reduced and zero-tillage are felt most strongly on medium-textured soils which are well drained and which possess a well developed structure. In some circumstances, compaction and associated poor drainage cannot be improved through good structural management and appropriate tillage practices alone, and installation of artificial drainage may be necessary.

Water Excess and Deficit

Excess water may be present in soils for a number of reasons. Compaction, for example, in the form of surface crusting, subsurface cultivation pans and hardsetting, often leads to reduced infiltration and restricted permeability. Excess water is also common in heavily textured clay soils, particularly in high rainfall areas, and in areas which are flat, low lying and prone to flooding. If excess water is found within about 40-50 cm of the surface, then the adverse effects on soil aeration and temperature are likely to lead to restricted root growth and crop performance.

In many cases, excess water may be removed from the soil through both sensible management of soil structure and appropriate tillage practices. Frequently, however, installation of artificial drainage is necessary. The aim of any artificial drainage programme is to lower the water table so that the amount of plant-available water in the rhizosphere is maximised. If the water table is lowered too much, the soil may become drought-susceptible.

Conversely, if the water table remains too high, waterlogging and poor aeration are likely to continue. The efficiency of drainage depends largely on the lateral spacing, depth and size of drains. Generally, efficiency improves with decreased lateral spacing, although at spacings of greater than around 6 m the rate of inflow to the drains is independent of spacing. Drain depth controls the height of the water table and usually occurs at about 50 cm to 2 m. In terms of size, the ability of a drain to carry water increases with increasing cross-sectional area.

The choice of drainage system depends on a number of factors, the most important being economic controls (installation and maintenance costs, expected yield improvements and availability of subsidies), management considerations (crop and tillage requirements), soil characteristics (texture, hydraulic conductivity and type of drainage problem) and climatic controls (rainfall amount, intensity and seasonality).

There are a number of specific drainage techniques which can be employed, either singly or in combination, depending on the type and extent of the drainage problem. Those most commonly used include subsoiling, mole drainage, pipe or tile drainage, open ditches or dykes, and regional or arterial drainage. Subsoiling is a form of deep ploughing and is often used to break up compacted cultivation pans and indurated layers within the soil.

The critical depth for this practice is about 20-50 cm; if it is shallower, then lifting of the soil may occur and if deeper, consolidation of the soil may result. Subsoiling is most effective when the soil is relatively dry and heavy; in wet conditions, plastic deformation and smearing may occur during passage of the subsoiling implement, and this is likely to have an adverse effect on soil structure.

Mole drainage is produced by drawing a mole plough, which is a bullet-shaped instrument usually less than 20 cm in diameter, through the soil to produce a continuous passage. Mole drains are established at a depth of about 40-60 cm, with a lateral spacing of about 2-5 m. They are not permanent features and are susceptible to collapse and sediment blockage. Consequently, they have to be renewed on a regular basis, usually after a period of about 5-7 years.

During the installation of mole drainage, the moisture status of the soil is of crucial importance. If conditions are too wet, smearing and consolidation may result, while shattering may occur if the soil is too dry. In spite of the temporary nature of mole drainage, it is often adopted in preference to other drainage practices because it is cheap and easy to install. It is also commonly used as a form of secondary drainage in combination with primary tile or ditch drains.

Pipe or tile drainage consists of plastic or ceramic pipes of about 10-30 cm in diameter, and is installed at a depth of 0.5-2.0 m. The pipes usually contain perforations or slits in order to speed up the transfer of water from the soil. Once the pipes have been installed, the ditches are often back-filled with gravel, polystyrene or fuel-ash. This helps to improve the flow of water to the drains, while acting as a filter to remove fine sediments which may otherwise lead to drain blockage.

Open ditches or dykes are usually constructed at field margins and are relatively cheap and easy to maintain. This type of drainage has been used since settled agriculture began, and in Britain dates back to the Roman period. In some circumstances, drainage may need to be improved at the regional scale by arterial drainage. This can be achieved in a number of ways, including straightening of river channels to increase gradient and improve flow, deepening of river channels by dredging, and pumping of water from agricultural land.

Flood prevention is also an important aspect of arterial drainage. In coastal areas, this is achieved through the construction of barrier systems which form part of sea defence strategies. Similarly, in low-lying floodplain areas, it may involve the construction of raised river banks or levées. Such regional or arterial drainage also has a long history, and large areas of land in many coastal, estuarine and low-lying riverine areas have been reclaimed in this way, such as the Fens, Somerset Levels and Romney Marsh in England.

One of the best-known examples of this type of large-scale reclamation is the Polder scheme in the Netherlands, where large areas were reclaimed from the sea in the low-lying coastal fringes. Since the Zuyder Zee (Reclamation) Act of 1918, four polders have been reclaimed, creating about 165,000 ha of new land. This involved the construction of large dykes which were fed by lateral drains spaced at 1.6-2.0 km intervals. These drains were supplemented by main drains spaced at 300 m intervals and then a coarse network of field drains spaced at 8-24 m intervals.

Although such large-scale drainage programmes have produced some of the world's most fertile and productive land, through improvements to a range of soil physical and chemical characteristics, they have also created problems. Perhaps the most notable of these is soil shrinkage, which is particularly common in areas dominated by marine clays and peats.

In parts of the Fens of eastern England, for example, it is estimated that the surface has been lowered by as much as 5 m following the long history of reclamation and drainage. In an area which is already low lying, surface lowering places even greater pressure on existing flood prevention and drainage schemes, many of which need to be upgraded. Also of concern in areas where large-scale reclamation and drainage are widespread, is the destruction of important wetland habitats, such as those in the Somerset Levels of southwest England.

Soil water deficit occurs most commonly in areas where rainfall is low in comparison with potential evapotranspiration. It is also common in coarse-textured, sandy soils which are excessively drained and low in organic matter, and in fine-textured, clay soils which contain large vertical cracks or which are poorly structured and therefore of low permeability; such soils have a low plant-available water capacity.

Soil water deficit does not necessarily occur throughout the year but is most common during warmer and drier periods. In Britain, for example, it rarely occurs in the October-March period but is common in the April-September period, especially in southern and eastern areas where it may occur in three to five years out of ten. Soil water deficit has the greatest impact where it occurs during the main period of crop growth, when it is demonstrated by excessive wilting of plants. In some circumstances, intense evapouration may lead to salinisation and sodification of soils. The main way in which soil water deficit is managed is by irrigation. This has been practised since the beginning of settled agriculture, not only in areas where rainfall is low throughout the year, but also where seasonal drought is a problem, in spite of respectable annual rainfall totals. It is used to replenish the soil moisture store and to leach toxic salts out of the rhizosphere, and as such it is a fundamental component of agriculture in many dryland areas.

The effectiveness of irrigation depends on the method and timing of water application, and on the amount and quality of water applied. There are three main methods—surface application, overhead application and subsurface application.

Surfacc application usually involves flood and furrow irrigation where water is diverted from existing river channels or canals directly onto the ridge and furrow systems of agricultural fields. This low cost and low technology approach is often preferred in developing countries, although it is also widely used in the developed world. One of the main problems with this method, however, is that capillary action may lead to redistribution of salts from the wet furrow areas to the drier ridges where the crops are often planted.

Overhead irrigation methods include spray or sprinkler, and trickle or drip systems. Spray irrigation systems involve the high pressure application of water through upright nozzles which may be attached to fixed or moveable pipes in a variety of ways, such as self-propelled water guns and centre-pivot systems.

Although water can be applied uniformly and efficiently using these methods, equipment costs may be prohibitive and evapourative losses are high, particularly on windy days. Moreover, soil erosion may occur as a result of overhead application, particularly in the early stages of crop growth when vegetation cover is low.

Trickle irrigation involves the application of water through flow regulating emitters which are placed at regular intervals in close proximity to the plants. In this way water is supplied more directly to the plant and evapourative losses are minimised.

Perhaps the most effective methods of irrigation involve subsurface applications of water. Here, ditches and underground drains or pipes are used and, although their installation can be costly, evapouration, soil structural damage and erosion losses are minimised.

In addition to irrigation, soil water deficit may also be managed through the control of evapouration from the soil surface. This is most commonly achieved through the addition of organic mulches, or through the establishment of a vegetation cover which can offer some degree of shade. These strategies are usually combined with sensible irrigation practices in order to increase water use efficiency in dry environments.

THE SOIL ATMOSPHERE

The irregularity of the soil particles in size shape, and arrangement insures the existence of a certain amount of space between them, even in the most tightly packed soils. This is termed the pore space of a soil. The pore space of soils varies from approximately 30 per cent of the volume of the soil in sandy soils to about so per cent of the volume in clay soils, or even higher in soils which are very rich in organic matter. The pore space of any given soil depends upon the physical and chemical conditions to which the soil is subjected. Conditions favouring a crumb structure of a soil, for example, usually result in an increase in its pore space. The interstitial spaces of a soil may be occupied entirely by air, as indesiccated soils, entirely by water, as in saturated soils, or, as is most usually true, partly by water and partly by air. The relative proportions of water and air present in any given soil vary depending upon the water content of the soil.

The air present in the soil is generally referred to as the soil atmosphere.The soil air usually contains a higher concentration of carbon dioxide and a lower concentration of oxygen than the above ground atmosphere. Concentrations of carbon dioxide as high as 5 per cent have been recorded for the soil atmosphere; such values are far in excess of the average value of 0.03 per cent for the air. The accumulation of carbon dioxide in soils is due more to the metabolic activities of micro-organisms than to respiration of soil animals and the underground portions of vascular plants. Except in very dry soils the soil atmosphere is usually saturated or nearly so with water-vapour. Soil Organisms. The soil flora includes bacteria, fungi, and algae. The bacteria are generally the most abundant of all the living organisms present in any soil. Among them are the nitrifying, sulfofying, nitrogen fixing, ammonifying, and cellulose decomposing bacteria. The bacteria accomplishing the oxidative decomposition of cellulose and similar compounds are the most important agents in the production of humus. The numbers of bacteria present vary greatly from soil to soil, and in any one soil vary with seasonal and other fluctuations in soil conditions. Most soils contain between two million and two hundred million individual bacteria per gram of soil. The number of bacteria decreases rapidly with increasing depth, sub soils being sterile or practically so.

In general an abundant representation of most species of bacteria is favoured in soils by warm temperatures (3 - 400 °C.), good aeration, and a good but not superabundant water supply. A high calcium content of the soil is also favourable to the development of most species of bacteria, apparently because it favours a granular structure of the soil, thus improving aeration. Some species of bacteria, on the other hand, are anaerobic,and thrive when aeration of the soil is deficient. The denitrifying bacteria and certain of the nitrogen-fixing bacteria (Clostridiurn spp.) are examples of anaerobes. Fungi are, in general, most abundant in soils of acid reaction. In such soils they largely replace bacteria as agents of decomposition of organic matter.

The soil fauna includes protozoa, nematodes, earthworms, insects, insect larvae, and burrowing species of the higher animals. The earthworms are generally credited with having the most important effects on soil structure, at least in many soils. Their activities result principally in a general loosening of the soil, which facilitates both aeration and distribution of water. Many of the other soil animals have similar effects on the structural organization of the soil.

SOIL WATER RELATIONS UNDER FIELD CONDITIONS

In a region of moist climate, if a hole be dug or bored into the ground at almost any place where the soil is deep enough a level at which the soil is completely saturated with water will be encountered at some depth or other. Water will stand in this hole up to this level of complete saturation which is called the water table. In river valleys or in close proximity to large bodies of water the water table will usually be reached at a depth of only a few feet under the soil surface. Even in regions of arid climate there are some local situations in which a water table is present.

The depth of the water table in any locality usually fluctuates, sometimes very markedly, with seasonal and periodic changes in the relative rates of rainfall, evapouration, transpiration and other factors. Relatively impermeable soil layers sometimes impede downward percolation of water sufficiently to cause the development of temporary water tables which may be far above the level of the true water table. Such temporary water tables have essentially the same effects on soil water relations as true water tables, except that their influence is often only a transient one.

For many years an important role was ascribed to the capillary rise of the water from the water table into the soil above in maintaining the moisture conditions within that soil. Recent investigations have shown, however, that the importance of this source of water in soils has been overemphasized. Experiments upon the rise of water through columns of soil indicate that water seldom rises through the soil by capillarity from the water table at an appreciable rate for heights of more than a few feet. In a typical loam soil the absolute maximum capillary rise of water is about 8 feet. Such capillary rise as does occur takes place most slowly, but to the greatest height in clay soils, and most rapidly but to lesser heights in sandy soils, loam soils occupying an intermediate position. In many soils the water table lies so far below the soil surface that it has little or no effect on the soil moisture conditions in the soil layers which are penetrated by the roots of most plants.

Generally speaking even in loam or clay loam soils, a capillary rise of water from a zone of complete saturation is probably ineffective in providing the roots of most species of plants with any considerable part of the water which they absorb unless the water table is within about 15 feet of the surface. Some of the more deeply rooted species may obtain some water from a water table located at depths as great as about 30 feet, but the presence of a water table at greater depths than this is an egligible factor in supplying the roots of any species of plants with water. A capillary rise of water from a water table is usually an important source of water for plants in such locations as river bottom lands or in fields or forests in close proximity to ponds or lakes. In many and perhaps most agricultural soils in moist climate regions plants obtain some water from the water table at least during the earlier part of the growing season. However the widespread practice of drainage of agricultural regions has often resulted in such a marked lowering of the water table as to greatly reduce the possibility that capillary rise of water from below will supply crop plants with any significant part of the water which they absorb. The other important source of soil water is that part of the precipitation usually rain, although often melting snow or ice, which percolates downward into the soil. In dry regions this is the only source of water available to plants.

Even in more humid regions this is the principal or only source of available soil water in most soils during the dry season of the year. Not all of the rain which falls on the surface of a soil penetrates into it. Some maybe lost by surface run-off and a portion usually evapourates before it can sink into the soil. The proportion of the precipitation which is lost in the run-off is in general much less with open, porous soils such as those found in most forests than with those of a lower porosity. A smaller proportion of the water is lost in the run-off when it falls as a gentle rain than when it comes as sudden downpour. The proportion of the precipitation which is lost by evapouration from the surface layers of the soil depends in part on the colour, texture porosity, and other properties of the soil, and in part on the temperature and vapour pressure of the atmosphere, as well as upon the impinging solar radiation. Let us now consider how percolating rainwater becomes distributed in as oil. We will assume that the soil under consideration is approximately air dry, that it is essentially homogeneous to a depth of several feet, and that Let us now assume that another two inches of rain falls on this same soil before any appreciable amount is lost from its upper layers by evapouration or transpiration. After a new equilibrium has been attained the upper foot of soil will not have increased in water content, but a layer of the soil approximately two feet in depth will now be moistened up to its field capacity. In other words addition of a second increment of water equal in volume to the first does not increase the water content of the already moist layer of the soil above its field capacity, but results, due to further capillary movement, in raising to its field capacity a second layer of soil, lying directly under the zone which had been moistened by the addition of the first increment of rain, and approximately equal to it in thickness.

Although the exact mechanics of the distribution of water under these conditions is probably more complicated, the effect produced is as if the second increment of water simply flows through the moist blanket of soil, after which it is distributed to the dry soil layer underneath by capillarity. Successive rainfalls would thus continue to deepen the layer of soil which has been moistened up to its field capacity. If the water table is not too deep, and if sufficient rainfall penetrates the soil, not too much of which is utilized by plants growing thereon, eventually the entire soil from its surface down to the water table will be moistened up to its field capacity. A zone several feet in height just above the water table may be enriched somewhat above field capacity by upward capillary movement from the water table.

If additional water is applied to the soil, after the entire soil mass down to the water table has attained its field capacity, this water percolates down through the soil under the influence of gravity and becomes apart of the ground water. Such conditions obtain in many of the soils of more humid regions at least during the wetter seasons of the year, provided that the water table is not located at too great a depth. This downward percolation of water to the water table is an important factor in influencing its depth below the soil surface, which usually fluctuates considerably from season to season.

In the preceding discussion we have assumed, in the interests of simplicity a homogeneous soil. However, most soils consist of a vertical succession of several distinct horizons, each with more or less distinctive physical and chemical properties. Even after long continued disturbance of a soil by plowing or other cultural activities at least some semblance of the original soil stratification usually persists. In the horizons below those reached by the plow the original structural organization of the soil is usually maintained practically intact. Although the individual horizons of a soil are often fairly homogeneous within themselves, each such horizon in a given soil may have a distinct field capacity of its own.

The water contents of the different horizons, even after an equilibrium in the capillary distribution of water has been attained, may therefore be very different. A somewhat erratic distribution of rainfall to an underlying soil often results from cracks or channels which are opened into the soil by one agency or another. Many soils crack upon drying, sometimes to depths of several feet. Such cracks often facilitate a mass flow of rainwater to a considerable depth in the soil. Similarly in forest soils the decomposition of roots may leave channels through which water flows down into the deeper layers of the soil. The burrows of animals may also facilitate the entry of water into a soil.The presence of rock strata, or of relatively impervious layers of soil close to the surface, will also modify very markedly the simple picture which has been presented of soil water conditions.

WATER RETAINING CAPACITY

One of the most familiar laboratory determinations of soil water is that of the so-called water retaining capacity (also called moisture holding capacity, water holding capacity, water capacity,etc.). This determination is usually made as follows: A shallow cylindrical pan with a perforated bottom is filled with dry soil, immersed in water to the depth of a few millimeters, and allowed to stand until it becomes saturated. The soil is then allowed to drain free of all the water which will drip out under the influence of the pull of gravity. The water content of the soil after it has been subjected to this treatment, determined by the usual method of oven-drying, is regarded as its water retaining capacity.

It has often been assumed that a determination such as that just described is an indication of the amount of water which would be held by that soil in the field in opposition to the pull of gravity but such an assumption is entirely unjustified. In the first place in making such determinations the vessel is filled with the soil in such a way that the distribution of particles in the soil mass is very different from that which obtains in the field.

This error is sometimes minimized by cutting out a core of soil in the field by means of a steel cylinder and using this core for a determination of the water retaining the plant. Since, no water is added to the soil the plants eventually pass into a state of permanent wilting due to the loss of water by transpiration. As soon as this occurs a sample of the soil is removed and its water content is determined by the oven-drying method. Prior to extensive determinations of wilting percentages it was supposed that plants differed very markedly in their capacity to reduce the water content of a soil.

It was assumed, for example, that species which could endure drought conditions could deplete the moisture content of a soil to a lower percentage before showing permanent wilting than could those species which were soon injured or killed when subjected to drought. Extensive investigations have shown, however, that hydrophytes, mesophytes, and xerophytes all reduce the water content of a given type of soil to about the same value before the condition of permanent wilting is induced.

The value of the wilting percentage for each species was determined by calculating the ratio of the individual determination to the mean of all determinations made with that soil. The values given in this table are the average mean ratios for a number of determinations on each species. While the wilting percentage for a given soil shows no appreciable variation when measured by means of different plants growing in it, the value varies greatly with the type of soil. The percentage of water remaining in a soil when permanent wilting of the plants growing in it occurs ranges from approximately 5 to 10 per cent for sandy loams, from about 10 to 15 per cent for silty loams, and from about 1 to 20 per cent in clay loams.

According to Caidwell (1913) and Shive and Livingston (1914) a greater percentage of water is left in the soil at permanent wilting if the plants are transpiring rapidly than if they are transpiring slowly. Veihmeyer and Hendrickson however, record that the wilting percentage of a given soil as determined with sunflower plants shows very little difference over a wide range of climatic conditions. In general this quantity seems to be controlled almost entirely by soil conditions and is only slightly influenced by the species of plant used, or by the climatic conditions to which that plant is exposed.

The significance of the wilting percentage lies in the fact that it is essentially a measure of that fraction of the soil water which is unavailable forplants. However, it should not be assumed that absolutely all movement of water into the roots has ceased when the soil water content has been reduced to its wilting percentage.

Wilted plants continue to reduce the water content of the soil but at such a slow rate that restoration of turgidity is impossible since, the rate of transpiration even from a wilted plant exceeds the rate of absorption of water from a soil which is at its wilting percentage or lower.

Actually, plants in the field often exhibit permanent wilting at soil water contents somewhat in excess of the wilting percentage as determined by the usual technique. In laboratory determinations of the wilting percentage the plants are ordinarily rooted in relatively small containers in which the soil mass becomes effectively penetrated by roots resulting in a fairly uniform reduction in the water content of the soil. In the field, however, such conditions do not often prevail.

The soil is usually less effectively penetrated by the root systems of the plant growing therein, and some portions of the soil may be less completely depleted of water than others before permanent wilting ensues. Although the soil in the immediate vicinity of the absorbing zone of each root is at the wilting percentage more remote portions of the soil may still be at the field capacity. Hence, the average water content of such a soil at permanent wilting may be considerably in excess of the wilting percentage as determined by the usual laboratory procedure.

2

The Mineral Component of the Soil

MINERAL COMPONENTS

Mineral particles are inorganic materials derived from rocks and minerals. They are extremely variable in size and composition.

COMPOSITION

Types of Primary and Secondary Minerals

Primary minerals are formed at high temperature and pressure, under reducing conditions without free oxygen. These minerals are mainly present in soils as sand and silt particles. They are not altered chemically since their deposition and crystallization is from molten lava.

Table. The Most Abundant Primary Minerals in Soils.

Mineral name	Formula
Quartz	(SiO_2)
Feldspars	($MAlSi_3O_8$ where M = Na, K, Ca)
Primary mica, such as muscovite	$KAl(AlSi_3O_{10})(OH)_2$
Primary mica, such as biotite	$K(Mg, Fe)(AlSiO_{10})(OH)_2$
Pyroxene	$MSiO_3$, where M = Mg, Mn
Amphiboles (*e.g.* mineral tremolite)	$Ca_2Mg_5Si_8O_{22}(OH)_2$
Olivine	$(Mg, Fe)_2SiO_4$

Most Common Primary Minerals in Soils

Secondary minerals are formed at low temperature and pressure through oxidation.

They are products of weathering of primary minerals either by alteration in the structure or re-precipitation. Usually secondary minerals are present in soil as clay particles.

Table. The Most Common Secondary Minerals in Soils.

Mineral name	Formula
Silicate clay minerals (or aluminosilicates)	One of the sub-groups are phyllosilicates.
Soluble minerals dolomite Ca, $Mg(CO_3)_2$	Carbonates, *e.g.* calcite $CaCO_3$ and Sulfates, *e.g.* gypsum $CaSO_4$ $2H_2O$.
Oxides or hydroxides of Al, Fe (sesquioxides) R_2O_3 nH_2O, where R = Al, Fe	Gibbsite $Al(OH)_3$ Goethite FeOOH Hematite Fe_2O_3.
Alophane Imogolite aluminosilicates (or clay minerals) Most Common Secondary Minerals in Soils	Amorphous (non-crystalline)

Most Common Secondary Minerals in Soils

Weathering refers to the breakdown and changes in rocks and sediments at or near the Earth's surface brought about by biological, chemical, and physical agents or combinations thereof.

Weathering also involves synthesis of new (secondary) minerals that are of a great importance in soil (*e.g.* clay minerals).

Examples of physical weathering include processes such as crystal growth, thermal expansion, moisture swelling, abrasion, etc. Chemical weathering consists of the following processes:

- *Hydration*: Intact water molecules bind to a mineral transforming hematite into ferrihydrate:
- *Hydrolysis*: Water molecules split into their hydrogen and hydroxyl components and hydrogen replaces a cation from the mineral structure (*e.g.* transformation of feldspar to kaolinite):
- *Dissolution (or Solution)*: Water is capable of dissolving many minerals by hydrating the cations and anions until they become dissociated from each other and surrounded by water molecules. (*e.g.* dissolution of gypsum):
- *Carbonation*: Weathering is accelerated by the presence of acids that increase the activity of hydrogen ions in water. For example, when carbon dioxide dissolves in water (a process enhanced by microbial and root respiration) the carbonic acid (H2CO3) produced hastens the chemical dissolution of calcite into limestone (or marble):
- *Oxidation-reduction*: Minerals that contain Fe, Mn, or S are especially susceptible to oxidation-reduction reactions. When for example, iron is oxidized from divalent to trivalent form, the change in valence and ionic radius causes destabilizing adjustments in the crystal structure of the mineral.

Biological weathering effects include the break up of rock particles by roots, the transfer and mixing of materials by burrowing animals,, and the formation of organo-mineral complexes (soil biological processes produce organic acids that can solubilize Al and Si ions, which are removed from a mineral by this process).

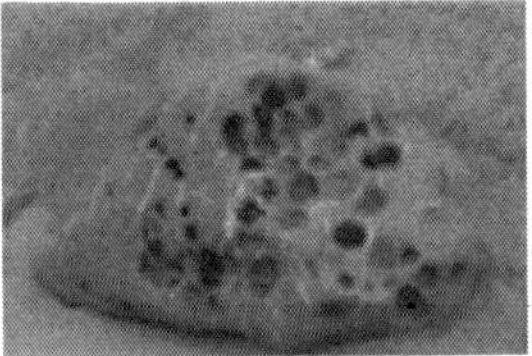

Fig. An example of biological weathering of a rock under the influence of clams.

Most Common Elements in Soils

Table. Contents of Some Elements in Soils, The Earth's Crust and Sediments.

Element	Soils (mg/kg)		Earthâ€s crust (mean)	Sediments (mean)
Â	Median	Range	Â	Â
O	490,000	-	474,000	486,000
Si	330,000	250,000-410,000	277,000	245,000
Al	71,000	10,000-300,000	82,000	72,000
Fe	40,000	2,000-550,000	41,000	41,000
C (total)	20,000	7,000-500,000	480	29,400
Ca	15,000	700-500,000	41,000	66,000
Mg	5,000	400-9,000	23,000	14,000
K	14,000	80-37,000	21,000	20,000
Na	5,000	150-25,000	23,000	5,700
Mn	1,000	20-10,000	950	770
Zn	90	1-900	75	95
Mo	1.2	0.1-40	1.5	2
Ni	50	2-750	80	52
Cu	30	2-250	50	33
N	2,000	200-5,000	25	470
P	800	35-5,300	1,000	670
S (total)	**700**	**30-1,600**	**260**	**2,200**

The median and range of various elements present in soils from around the world are given in Table. The elements that are found in soils in the highest quantities are O, Si, Al, Fe, C, Ca, K, Na, and Mg. These are also major elements found in the Earth's crust and in sediments. Oxygen is the most prevalent element in the Earth's crust and in soils. It comprises about 47 per cent of the Earth's crust by weight and more than 90 per cent by volume.

Particle and Bulk Density

- Particle density (r_s) is mass of solids.
- (M_s) per volume of solids (V_s).

$$r_s = M_s/V_s$$

In most mineral soils the mean density of the particles is about 2.6-2.7 gm/cm^3 (or 2600 - 2700 kg/m^3). Soils with a high content of iron oxides and various heavy minerals have particle density of 5.2-5.3 gm/cm^3, while soils with high organic matter content can have particle density as low as 1.3 gm/cm^3.

Bulk density (r_b) is the mass of solids (M_s) per total soil volume (V_t).

$$r_b = M_s/ V_t$$

$$r_b = M_s/ (V_s + V_a + V_w)$$

where:

V_s=volume of solids;
V_a=volume of air;
V_w=volume of water

Bulk density is always smaller thanr_s. Since in a general case pores constitute half the volume, r_b is about half ofr_s, namely 1.3-1.35 g/cm^3 (or 1300-1350 kg/m^3).

Size of Soil Particles

The mineral particles present in soils vary enormously in size from boulders and stones down to sand grains and minute clay particles that cannot be seen by an optical microscope. An arbitrary division is made by size-grading soil into material:

- That passes trough a sieve with 2-mm diameter holes the fine earth (consisting of sand, silt, and clay particles),
- That is retained on the sieve (> 2 mm) - the coarse fragments (gravel, cobbles, and stones).

Coarse fragments (diameter > 2 mm) are defined as rock fragments and do not include fragments of pads or concretions.

Soil Texture

Soil texture refers to the relative proportions of sand, silt, and clay in a soil. It is often the first and most important property to be determined when describing a soil, since many conclusions can be drawn from this information (water intake or infiltration, water storage in the soil, soil aeration, soil fertility, trafficability, etc.).

A number of different classification systems have been devised to divide soil particles according to their size. The figure below shows the limits of different size classes in three systems of soil particle size classification. Although Canadian classification system is not shown, it is very similar to the USDA system, with an exception of sub-classes of silt and clay size classes. Textural names are given to soils based on the relative proportions of each of the three soil particles - sand, silt, and clay. Soils that contain predominantly clay size particles are called clay (textural class), those with high silt content are called silt (textural class), and those with a high sand content are called sand (textural class). A more complicated soil textural class that is called a loam is a mixture of sand, silt, and clay particles and it exhibits the properties of those particles in about equal proportions.

Soil colloids

Types of Soil Colloids

When suspended in water, colloids do not readily settle out due to Brownian motion (when particles are moved around by collisions with other small particles). In chemistry colloids are particles smaller than 0.001 mm (or 1 mm), but it has been observed that soil particles smaller than 0.002 mm (or 2 mm) still behave in the same manner, hence they are called soil colloids. Most common soil colloids are:

- Phyllosilicate clay minerals,
- Oxides and hydrous oxides of iron and aluminum (*e.g.*, gibbsite, goethite, hematite),
- Amorphous minerals (*e.g.*, allophane and imogolite),
- Organic colloids.

Phyllosilicate clay minerals are a sub-group of silicates (an extensive group of minerals which are derived from silica SiO2). All silicates are based on the SiO44- tetrahedron with partial or complete sharing of oxygen atoms. Different silicate structures arise from the various ways in which SiO44- tetrahedra combine with one another. Two types of building blocks are present in phyllosilicates:

- Tetrahedron (4-faced structure) where each silicon (Si) is surrounded with 4 oxygen atoms.
- Octahedron (8-faced structure) where each aluminum atom is surrounded by 6 oxygen atoms.

Two types of phyllosilicate clay minerals that are commonly present in soils are montmorillonite and kaolinite. Montmorillonite is a 2:1 phyllosilicate mineral where there is one Al-octahedral sheet sandwiched between two Si-tetrahedral sheets (one such sandwich is called a layer, while many layers make a micelle or crystal). The layers in montmorillonite crystals are loosely held together by very weak oxygen-to-oxygen and cation-to-oxygen bonds. Hence, water molecules and various ions are attracted between layers (into the interlayer space) causing expansion of the crystal. Montmorillonite is characterized by pronounced swelling when wet and shrinkage on drying. Wide cracks commonly form as montmorillonite dominated soils (*e.g.* Vertisols) are dried. Typically montmorillonite particles are thin (thickness ~1 to 3 nm and width ~500 nm), flat disks.

Kaolinite is a 1:1 phyllosilicate mineral in which there is one Al-octahedral sheet bonded to one Si-tetrahedral sheet by shared (apical) oxygen atoms of the tetrahedral sheet. Two adjacent layers are held together by hydrogen bonding. Consequently, the structure is fixed and no expansion occurs between layers when the clay is wetted. The effective surface of kaolinite is restricted to its external surface area. The typical kaolinite particle is fairly large (thickness ~ 50 nm and width ~ 1-3 nm) in comparison with other clay particles and hexagonal in shape. Because of the strong binding forces between their structural layers, kaolinite particles are not readily broken down into extremely thin plates. Kaolinite exhibits very little shrinkage and swelling.

Amorphous Minerals (Imogolite and Allophane)

Imogolite and allophane are minerals found in many soils including soils formed on volcanic ash parent material and Podzolic soil type. They are very small (about 2 nanometers in diameter) and a very reactive component of soils.

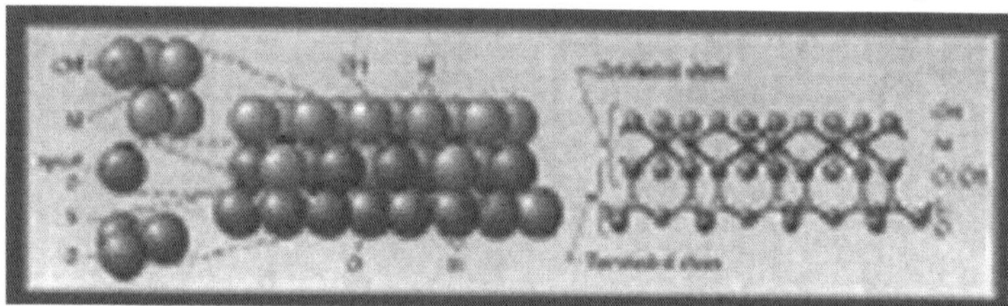

Imogolite and allophane are amorphous, which means that they do not have a repetitive structure with atoms arranged in a regular pattern over relatively long distances (at least tens of nanometers). Because their structure is somewhat ordered on a few nanometers scale, they are said to be 'paracrystalline' or to have 'short range order'.

Imogolite and allophane are both aluminosilicates, which means that they are made of variable amounts of aluminum (Al), silicon (Si), oxygen (O), and hydrogen (H) ions. They have a rather similar composition but their morphology is different. Imogolite forms tubes while allophane forms hollow spheres.

Specific Surface Area

Specific surface area is defined as surface area of particles per unit mass or unit volume of particles. Because of their large size sand particles have small specific surface area (*e.g.*, 0.1 m^2/g).

Clay particles, on the other hand, have very large specific surface area, giving them a tremendous capacity to adsorb water and other substances (*e.g.* 10-1,000 m^2/g).

Charges on Soil Colloids

"Ideal" montmorillonite composition is HSi_2AlO_6, which results in an "ideal" net charge of [1(+1)+2(+4)+1(+3)+6(-2)] = 0. However, the real composition of montmorillonite involves isomorphic substitution of Mg for some Al in the octahedral sheet. With Mg^{2+} substituting for Al^{3+} there are fewer positive charges to neutralize the negative charges and a large permanent (negative) charge results (it is called permanent charge since it takes place during crystallization and it is not a subject to changes). In contrast, there are also variable or pH-dependent charges that are formed by protonation and deprotonation of functional groups like -OH.

"Ideal" kaolinite composition is $H_4Si_2Al_2O_9$, which results in an "ideal" net charge of [4(+1)+2(+4)+2(+3)+9(-2)] = 0. Composition of kaolinite is close to ideal, since there is very little or no isomorphic substitution, resulting in no permanent charges. Kaolinite in soil usually has some surface contamination, and net negative charge is about 8 $cmol_c$/kg soil. In addition, kaolinite has a small amount of variable charge associated with dissociable OH groups attached to Al at the edges of the crystal, where the structural regularity ends.

3

Nutrient Cycling

INTRODUCTION

In ecology, a biogeochemical cycle/ nutrient cycle is a circuit or pathway by which a chemical element or molecule moves through both biotic ("bio-") and abiotic ("geo-") compartments of an ecosystem. In effect, the element is recycled, although in some such cycles there may be places (called "sinks") where the element is accumulated or held for a long period of time.

All chemical elements occurring in organisms are part of biogeochemical cycles. In addition to being a part of living organisms, these chemical elements also cycle through abiotic factors of ecosystems such as water (hydrosphere), land (lithosphere), and the air (atmosphere); the living factors of the planet can be referred to collectively as the biosphere.

All the chemicals, nutrients, or elements—such as carbon, nitrogen, oxygen, phosphorus—used in ecosystems by living organisms operate on a closed system, which refers to the fact that these chemicals are recycled instead of being lost and replenished constantly such as in an open system. The energy of an ecosystem occurs on an open system; the sun constantly gives the planet energy in the form of light while it is eventually used and lost in the form of heat throughout the trophic levels of a food web.

The Earth does not constantly receive more chemicals as it receives light. The Earth only has those chemicals that were formed in the creation of the Earth, and the only way to obtain more chemicals or nutrients is from occasional meteorites from outer space that contain those elements. Because chemicals operate on a closed system and cannot be lost and replenished like energy can, these chemicals must be recycled throughout all of Earth's processes that use those chemicals or elements.

These cycles include both the living biosphere, and the non-living lithosphere, atmosphere, and hydrosphere. The term "biogeochemical" takes its prefixes from these cycles: Bio refers to the biosphere. Geo refers collectively to the lithosphere, atmosphere, and hydrosphere. Chemical, of course, refers to the chemicals that go through the cycle.

The chemicals are sometimes held for long periods of time in one place. This place is called a reservoir, which, for example, includes such things as coal deposits that are storing carbon for a long period of time.

When chemicals are held for only short periods of time, they are being held in exchange pools. Generally, reservoirs are abiotic factors while exchange pools are biotic factors. Examples of exchange pools include plants and animals, which temporarily use carbon in their systems and release it back into the air or wherever. Carbon is held for a relatively short time in plants and animals when compared to coal deposits. The amount of time that a chemical is held in one place is called its residence.

The most well-known and important biogeochemical cycles, for example, include the carbon cycle, the nitrogen cycle, the oxygen cycle, the phosphorus cycle, and the water cycle.

Biogeochemical cycles always involve equilibrium states: a balance in the cycling of the element between compartments. However, overall balance may involve compartments distributed on a global scale. Biogeochemical cycles of particular interest in ecology are:

- Nitrogen cycle,
- Oxygen cycle,
- Carbon cycle,
- Phosphorus cycle,
- Sulphur cycle,
- Water cycle,
- Hydrogen cycle.

NITROGEN CYCLE

The nitrogen cycle is a much more complicated biogeochemical cycle but also cycles through living parts and non-living parts including the water, land, and air. Nitrogen is a very important molecule in that it is part of both proteins, present in the composition of the amino acids that make up proteins, as well as nucleic acids such as DNA and RNA, present in nitrogenous bases. The largest reservoir of nitrogen is the atmosphere, in which about 78 per cent of nitrogen is contained as nitrogen gas (N2). Nitrogen gas is "fixed," in a process called nitrogen fixation. Nitrogen fixation combines nitrogen with oxygen to create nitrates (NO3).

Nitrates can then be used by plants or animals (which eat plants or eat animals that have eaten plants). Nitrogen can be fixed either by lightning, industrial methods (such as for fertilizer), in free nitrogen-fixing bacteria in the soil, as well as in nitrogen-fixing bacteria present in roots of legumes (such as rhizobium). Nitrogen-fixing bacteria use certain enzymes that are capable of fixing nitrogen gas into nitrates and include free bacteria in soil, symbiotic bacteria in legumes, and also cyanobacteria, or blue-green algae, in water.

After being used by plants and animals, nitrogen is then disposed of in decay and wastes. Detritivores and decomposers decompose the detritius from plants and animals, nitrogen is changed into ammonia, or nitrogen with 3 hydrogen atoms (NH3). Ammonia is toxic and cannot be used by plants or animals, but nitrite bacteria present in the soil can take ammonia and turn it into nitrite, nitrogen with two oxygen atoms (NO2). Although nitrite is also unusable by most plants and animals, nitrate bacteria changes nitrites back into nitrates, usable by plants and animals. Some nitrates are also converted back into nitrogen gas through the process of denitrification, which is the opposite of nitrogen-fixing, also called nitrification. Certain denitrifying bacteria are responsible for this.

OXYGEN CYCLE

The oxygen cycle is the biogeochemical cycle that describes the movement of oxygen within and between its three main reservoirs: the atmosphere, the biosphere, and the lithosphere. The main driving factor of the oxygen cycle is photosynthesis, which is responsible for the modern Earth's atmosphere and life as we know it. If all photosynthesis were to cease, the Earth's atmosphere would be devoid of all but trace amounts of oxygen within 5000 years. The oxygen cycle would no longer exist.

CARBON CYCLE

The carbon cycle is the biogeochemical cycle by which carbon is exchanged between the biosphere, geosphere, hydrosphere and atmosphere of the Earth. Other bodies may have carbon cycles, but little is known about them. All of these components are reservoirs of carbon. The cycle is usually thought of as four main reservoirs of carbon interconnected by pathways of exchange. The reservoirs are the atmosphere, terrestrial biosphere (usually includes freshwater systems), oceans, and sediments (includes fossil fuels). The annual movements of carbon, the carbon exchanges between reservoirs, occur because of various chemical, physical, geological, and biological processes. The ocean contains the largest pool of carbon near the surface of the Earth, but most of that pool is not involved with rapid exchange with the atmosphere.

The global carbon budget is the balance of the exchanges (incomes and losses) of carbon between the carbon reservoirs or between one specific loop (*e.g.*, atmosphere - biosphere) of the carbon cycle. An examination of the carbon budget of a pool or reservoir can provide information about whether the pool or reservoir is functioning as a source or sink for carbon dioxide.

PHOSPHORUS CYCLE

The phosphorus cycle is the biogeochemical cycle that describes the movement of phosphorus through the lithosphere, hydrosphere, and biosphere.

Unlike many other biogeochemicals, the atmosphere does not play a significant role in the movements of phosphorus, because phosphorus and phosphorus-based compounds are usually solids at the typical ranges of temperature and pressure found on Earth.

SULPHUR CYCLE

Sulphur is one of the constituents of many proteins and vitamins and hormones. It recycles like other biogeochemical cycles. The essential steps of the sulphur cycle are:

- Mineralization of organic sulphur to the inorganic form, hydrogen sulfide: (H_2S).
- Oxidation of sulfide and elemental sulphur (S) and related compounds to sulfate (SO_4^{2-}).
- Reduction of sulfate to sulfide.
- Microbial immobilization of the sulphur compounds and subsequent incorporation into the organic form of sulphur.

WATER CYCLE

The water cycle — technically known as the hydrologic cycle — is the continuous circulation of water within the Earth's hydrosphere, and is driven by solar radiation. This includes the atmosphere, land, surface water and groundwater.

As water moves through the cycle, it changes state between liquid, solid, and gas phases. Water moves from compartment to compartment, such as from river to ocean, by the physical processes of evaporation, precipitation, infiltration, run-off, and subsurface flow. Movement of water within the water cycle is the subject of the field of hydrology.

HYDROGEN CYCLE

Hydrogen is one of the constituents of water. It recycles as in other biogeochemical cycles. It is actively involved with the other cycles like the carbon cycle, nitrogen cycle and sulphur cycle as well.

Anaerobic decomposition of organic substances to carbon dioxide and methane is a collaborative effort involving many different reactions and species of microorganisms. This is also called *interspecies hydrogen transfer*.

NUTRIENTS FOR PLANT GROWTH

Plant nutrition often is confused with fertilization. Plant nutrition refers to a plant's need for and use of basic chemical elements. Fertilization is the term used when these materials are added to the environment around a plant. A lot must happen before a chemical element in a fertilizer can be used by a plant. Plants need 17 elements for normal growth. Three of them—carbon, hydrogen, and oxygen—are found in air and water. The rest are found in the soil. Six soil elements are called macronutrients because they are used in relatively large amounts by plants. They are nitrogen, potassium, magnesium, calcium, phosphorus, and sulphur in Table.

Table. Plant Macronutrients.

Element	Absorbed as	Leaches from soil/ Mobility in plant	Signs of excess	Signs of deficiency	Notes
Nitrogen (N)	NO_3^-, (nitrate), NH_4^+ (ammonium)	Leachable, especially NO_3^-. Mobile in plants.	Succulent growth; dark green color; weak, spindly growth; few fruits. May cause brittle growth, especially under high temperatures.	Reduced growth, yellowing (chlorosis). Reds and purples may intensify in some plants. Reduced lateral bud breaks. Symptoms appear first on older growth.	In general, the best NH_4^+:NO_3^- ratio is 1:1. Under low sugar conditions (low light), high NH_4^+ can cause leaf curl. Uptake is inhibited by high P levels. The N:K ratio is extremely important. Indoors, the best N:K ratio is 1:1 unless light is extremely high. In soils with a high C:N ratio, more N should be supplied.
Phosphorus (P)	$H_2PO_4^-$, HPO_4^- (phosphate)	Normally not leachable, but may leach from soil high in bark or peat. Not readily mobile in plants.	Shows up as micronutrient deficiency of Zn, Fe, or Co.	Reduced growth. Color may intensify; browning or purpling of foliage in some plants. Thin stems, reduced lateral bud breaks, loss of lower leaves, reduced flowering.	Rapidly bound (fixed) on soil (P) particles. Under acid conditions, fixed with Fe, Mg, and Al. Under alkaline conditions, fixed with Ca. Important for young plant and seedling growth. High P interferes with micronutrient absorption and N absorption. Used in relatively small amounts when compared to N and K.
Potassium (K)	K^+	Can leach in sandy soils. Mobile in plants.	Causes N deficiency in plant and may affect the uptake of other positive ions.	Reduced growth, shortened inter-nodes. Marginal burn or scorch (brown leaf edges), necrotic (dead) spots in leaves. Reduction of lateral bud breaks, tendency to wilt readily.	N:K balance is important. High N:low K favors vegetative growth; low N:high K promotes reproductive growth (flowers, fruit).

Eight other soil elements are used in much smaller amounts and are called micronutrients or trace elements in Table.

Table. Plant Micronutrients.

Element	Absorbed as	Signs of excess	Signs of deficiency	Notes
Iron (Fe)	Fe++, Fe+++	Rare except on flooded soils. Interveinal chlorosis, primarily on young tissue, which eventually may turn white.	Soil high in Ca, Mn, P, or heavy metals (Cu, Zn); high pH; poorly drained soil; oxygen-deficient soil; nematode attack on roots.	Add Fe in the chelate form. The type of chelate needed depends on soil pH.
Boron (B)	BO3- Borate	Blackening or death of tissue between veins.	Failure to set seed, internal breakdown, death of apical buds.	
Zinc (Zn)	Zn++	Shows up as Fe deficiency. Also interferes with Mg absorption.	"Little leaf" (reduction in leaf size), short internodes, distorted or puckered leaf margins, interveinal chlorosis.	
Copper (Cu)	Cu++, Cu+	Can occur at low pH. Shows up as Fe deficiency.	New growth small, mis-shapen, wilted.	May be found in some peat soils.
Manganese (Mn)	Mn++	Reduction in growth, brown spotting on leaves. Shows up as Fe deficiency.	Interveinal chlorosis of leaves followed by brown spots, producing a checkered effect.	Found under acid conditions.
Molybdenum (Mo)	MoO4- (molybdate)		Interveinal chlorosis on older or midstem leaves, twisted leaves (whiptail).	
Chlorine (Cl)	Cl-	Salt injury, leaf burn. May increase succulence.	Leaves wilt, then become bronze, then chlorotic, then die; club roots.	

They are iron, zinc, molybdenum, manganese, boron, copper, cobalt, and chlorine. Most of the nutrients a plant needs are dissolved in water and then absorbed by its roots. In fact, 98 per cent are absorbed from the soil-water solution, and only about 2 per cent are actually extracted from soil particles.

FERTILIZERS

Fertilizers are materials containing plant nutrients that are added to the environment around a plant. Generally, they are added to the water or soil, but some can be sprayed on leaves. This method is called foliar fertilization. It should be done carefully with a dilute solution, because a high fertilizer concentration can injure leaf cells. The nutrient, however, does need to pass through the thin layer of wax (cutin) on the leaf surface.

Fertilizers are not plant food! Plants produce their own food from water, carbon dioxide, and solar energy through photosynthesis. This food (sugars and carbohydrates) is combined with plant nutrients to produce proteins, enzymes, vitamins, and other elements essential to growth.

NUTRIENT ABSORPTION

Anything that reduces or stops sugar production in leaves can lower nutrient absorption. Thus, if a plant is under stress because of low light or extreme temperatures, nutrient deficiency may develop.

A plant's developmental stage or rate of growth also may affect the amount of nutrients absorbed. Many plants have a rest (dormant) period during part of the year. During this time, few nutrients are absorbed. Plants also may absorb different nutrients as flower buds begin to develop than they do during periods of rapid vegetative growth.

A nutrient is any element or compound necessary for or contributing to an organism's metabolism, growth, or other functioning. Six nutrient groups exist, classifiable as those that provide energy, and as those that otherwise support metabolic processes in the body: Some of them are essential because they cannot be synthesized in the body and must be obtained from a food source.

SUBSTANCES THAT PROVIDE ENERGY

Carbohydrates are compounds made up of sugars. Carbohydrates are classified by their number of sugar units: monosaccharides (such as glucose and fructose), disaccharides (such as sucrose and lactose), oligosaccharides, and polysaccharides (such as starch, glycogen, and cellulose).

Proteins are organic compounds that consists of amino acids joined by peptide bonds. The body does not manufacture certain amino acids (termed essential amino acids); the diet must supply these. In nutrition, proteins are broken down through digestion back into free amino acids.

Fats consist of a glycerin molecule with three fatty acids attached. Fatty acids are unbranched hydrocarbon chains, connected by single bonds alone (saturated fatty acids) or by both single and double bonds (unsaturated fatty acids). Fats are needed to keep cells functioning properly, to insulate body organs against shock, to keep body temperature stable, and to maintain healthy skin and hair. The body does not manufacture certain fatty acids (termed essential fatty acids); the diet must supply these.

Fat has an energy content of 9 kcal/g (~37.7 kJ/g); proteins and carbohydrates 4 kcal/g (~16.7 kJ/g). Ethanol (grain alcohol) has an energy content of 7 kcal/g (~29.3 kJ/g).

SUBSTANCES THAT SUPPORT METABOLISM

Minerals are generally trace elements, salts, or ions such as copper and iron. These minerals are essential to human metabolism.

Vitamins are organic compounds essential to the body. They usually act as coenzymes for various proteins in the body.

Water is an essential nutrient and is directly involved in all the chemical reactions of life.[citation needed]

NUTRITION AS A SCIENCE

Any classification of "nutrients" is likely to be arbitrary given the status of nutrition as a developing science. Researchers are becoming more aware of a wider range of nutrients essential for health.

An organism will metabolise any organic compound to use for its energy content, for structural purposes (growth or replacement of living structures), or for participation in chemical reactions necessary for life. Any particular substance can play more than one role in the body, though researchers lack a good understanding of these roles.

The discovery of the group of nutrients called phytonutrients reinforces the provisional nature of our knowledge. We know little about phytonutrients, organic compounds from plants, which play an essential role in the normal functioning of a body and have complex hormonal effects on health, or play an active role in the amelioration of disease. They do not fit readily into the scheme of the traditional nutrition categories.

NUTRIENTS AND THE ENVIRONMENT

The band of a green alga (*Enteromorpha*) along this shore indicates that there is a nearby source of nutrients (probably nitrates or ammonia from a small estuary).

While in essence true to the definition above, the term nutrients has a more limited meaning within the specialised fields of water quality and water pollution, referring specifically to plant fertilizers. In this context, certain mineral compounds can have an adverse impact on water quality because of their ability to promote plant and algae growth. An excessive growth of aquatic plants can clog waterways, and over-stimulation of algae and microbes leads to an ecological process called eutrophication.

A surprisingly small number of elements provide interest or concern in this context: really just nitrogen and phosphorus in most aquatic systems. Mineral compounds involved are ammonia, nitrites, nitrates, and orthophosphates. Organic compounds also may contribute, in as much as they also contain nitrogen and phosphorus. The reason only a few chemicals are of concern has to do with the fact that plants are made up mostly of compounds of carbon (C), hydrogen (H), oxygen (O), nitrogen (N), and phosphorus (P), and lesser amounts of sulphur (S), potassium (K), magnesium (Mg), and calcium (Ca). These elements constitute the macronutrients. Many other elements, though necessary for growth, classify as micronutrients due to the very small quantities required.

Plants obtain carbon, hydrogen, and oxygen (elements most needed for growth) from the air and water, where all three elements occur in great abundance as water and as carbon dioxide. Nutrients having greatest potential to influence plant growth in aquatic environments is those elements needed for plant growth in proportionately large amounts (that is, macronutrients) but likely to become limiting—that is, present in amounts that could be depleted by continued growth. Once used up, further growth will not be possible. Of the nine macronutrients, nitrogen and phosphorus are most likely to become limiting.

The others always remain present in great abundance (C, H, O) or usually in amounts that exceed the requirements of aquatic plants or algae.

Farmers apply fertilizer nutrients in the form of nitrogen, phosphorus, and potassium (N, P, and K with perhaps micronutrients) to prevent these elements from becoming limiting in the soil. These elements become concentrated in wastewaters from animal pens and septic or sewage systems. And these elements (especially N and P) in run-off or wastewater discharges reaching streams, lakes, or seas will promote aquatic plant growth. Abundant plant growth itself gives cause for concern in assessing water quality. The most abundant "plants" in most aquatic environments are algae. When essential nutrients are plentiful, algae multiply. If these algae are microscopic phytoplankton, their growth increases the turbidity of the water. The water then becomes cloudy, coloured a shade of green, yellow, or brown (sometimes red). A super abundance of algae, or of higher plants, in an aquatic system can signal excessive inputs of nutrients.

THE PATHWAY OF NITROGEN

ROLE OF NITROGEN IN THE BIOSPHERE

The growth of all organisms depends on the availability of mineral nutrients, and none is more important than nitrogen, which is required in large amounts as an essential component of proteins, nucleic acids and other cellular constituents.

There is an abundant supply of nitrogen in the earth's atmosphere - nearly 79 per cent in the form of N2 gas. However, N2 is unavailable for use by most organisms because there is a triple bond between the two nitrogen atoms, making the molecule almost inert. In order for nitrogen to be used for growth it must be "fixed" (combined) in the form of ammonium (NH4) or nitrate (NO3) ions. The weathering of rocks releases these ions so slowly that it has a neglible effect on the availability of fixed nitrogen. So, nitrogen is often the limiting factor for growth and biomass production in all environments where there is suitable climate and availability of water to support life.

Microorganisms have a central role in almost all aspects of nitrogen availability and thus for life support on earth:

Some bacteria can convert N2 into ammonia by the process termed nitrogen fixation; these bacteria are either free-living or form symbiotic associations with plants or other organisms (*e.g.* termites, protozoa) other bacteria bring about transformations of ammonia to nitrate, and of nitrate to N2 or other nitrogen gases many bacteria and fungi degrade organic matter, releasing fixed nitrogen for reuse by other organisms.

All these processes contribute to the nitrogen cycle.

We shall deal first with the process of nitrogen fixation and the nitrogen-fixing organisms, then consider the microbial processes involved in the cycling of nitrogen in the biosphere.

NITROGEN FIXATION

A relatively small amount of ammonia is produced by lightning. Some ammonia also is produced industrially by the Haber-Bosch process, using an iron-based catalyst, very high pressures and fairly high temperature. But the major conversion of N2 into ammonia, and thence into proteins, is achieved by microorganisms in the process called nitrogen fixation (or dinitrogen fixation).

Table. Type of Fixation N2 Fixed (1012 g per year, or 106 metric tons per year).

Type of Fixation	N_2 fixed (10^{12} g per year, or 10^6 metric tons per year)
Non-biological	Â>
Non-biological	Â
Industrial	about 50
Combustion	about 20
Lightning	about 10
Total	about 80
Â	Â
Biological	Â
Agricultural land	about 90
Forest and non-agricultural land	about 50
Sea	about 35
Total	about 175

The table above shows some estimates of the amount of nitrogen fixed on a global scale. The total biological nitrogen fixation is estimated to be twice as much as the total nitrogen fixation by non-biological processes. To illustrate the importance of biological nitrogen fixation, the image below shows part of the Lower Sonoran desert in Arizona.

Every plant in this scene depends ultimately on biological nitrogen fixation. Both free-living cyanobacteria and the cyanobacterial associates of lichens initially contributed nitrogen to the soil by forming a cryptobiotic crust. Now numerous leguminous plants occur in this desert, with nitrogen-fixing Rhizobium in their root nodules. Examples are the green-stemmed brush-like trees at the right and left of the image (Parkinsonia species, common name "paloverde"), and several acacias and mesquites.

Mechanism of Biological Nitrogen Fixation

Biological nitrogen fixation can be represented by the following equation, in which two moles of ammonia are produced from one mole of nitrogen gas, at the expense of 16 moles of ATP and a supply of electrons and protons (hydrogen ions):

$$N_2 + 8H+ + 8e- + 16\,ATP = 2NH_3 + H_2 + 16ADP + 16\,Pi$$

This reaction is performed exclusively by prokaryotes (the bacteria and related organisms), using an enzyme complex termed nitrogenase. This enzyme consists of two proteins - an iron protein and a molybdenum-iron protein, as shown below.

The reactions occur while N2 is bound to the nitrogenase enzyme complex. The Fe protein is first reduced by electrons donated by ferredoxin. Then the reduced Fe protein binds ATP and reduces the molybdenum-iron protein, which donates electrons to N2, producing HN=NH. In two further cycles of this process (each requiring electrons donated by ferredoxin) HN=NH is reduced to H2N-NH2, and this in turn is reduced to 2NH3.

Depending on the type of microorganism, the reduced ferredoxin which supplies electrons for this process is generated by photosynthesis, respiration or fermentation.

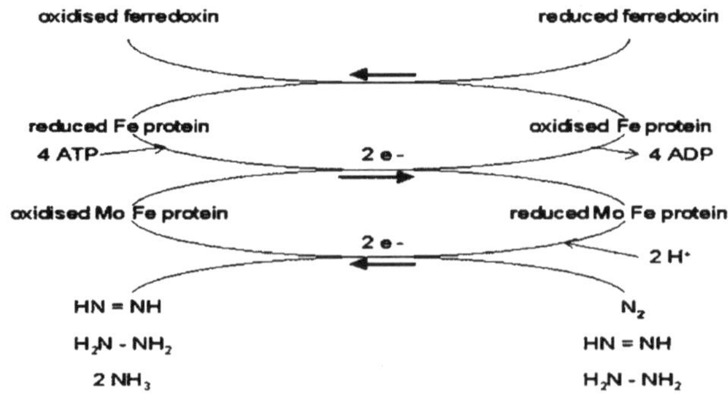

There is a remarkable degree of functional conservation between the nitrogenase proteins of all nitrogen-fixing bacteria. The Fe protein and the Mo-

Fe protein have been isolated from many of these bacteria, and nitrogen fixation can be shown to occur in cell-free systems in a laboratory when the Fe protein of one species is mixed with the Mo-Fe protein of another bacterium, even if the species are very distantly related.

The Nitrogen-fixing Organisms

All the nitrogen-fixing organisms are prokaryotes (bacteria). Some of them live independently of other organisms - the so-called free-living nitrogen-fixing bacteria. Others live in intimate symbiotic associations with plants or with other organisms (*e.g.* protozoa). Examples are shown in the table below.

A point of special interest is that the nitrogenase enzyme complex is highly sensitive to oxygen. It is inactivated when exposed to oxygen, because this reacts with the iron component of the proteins. Although this is not a problem for anaerobic bacteria, it could be a major problem for the aerobic species such as cyanobacteria (which generate oxygen during photosynthesis) and the free-living aerobic bacteria of soils, such as Azotobacter and Beijerinckia. These organisms have various methods to overcome the problem. For example, Azotobacter species have the highest known rate of respiratory metabolism of any organism, so they might protect the enzyme by maintaining a very low level of oxygen in their cells. Azotobacter species also produce copious amounts of extracellular polysaccharide (as do Rhizobium species in culture -). By maintaining water within the polysaccharide slime layer, these bacteria can limit the diffusion rate of oxygen to the cells. In the symbiotic nitrogen-fixing organisms such as Rhizobium, the root nodules can contain oxygen-scavenging molecules such as leghaemoglobin, which shows as a pink colour when the active nitrogen-fixing nodules of legume roots are cut open. Leghaemoglobin may regulate the supply of oxygen to the nodule tissues in the same way as haemoglobin regulates the supply of oxygen to mammalian tissues. Some of the cyanobacteria have yet another mechanism for protecting nitrogenase: nitrogen fixation occurs in special cells (heterocysts) which possess only photosystem I (used to generate ATP by light-mediated reactions) whereas the other cells have both photosystem I and photosystem II (which generates oxygen when light energy is used to split water to supply H2 for synthesis of organic compounds).

Symbiotic Nitrogen Fixation

The most familiar examples of nitrogen-fixing symbioses are the root nodules of legumes (peas, beans, clover, etc.). Part of a clover root system bearing naturally occurring nodules of Rhizobium. Each nodule is about 2-3 mm long. Clover root nodules at higher magnification, showing two partly crushed nodules (arrowheads) with pink-coloured contents. This colour is caused by the presence of the pigment leghaemoglobin - a unique metabolite of this type of symbiosis. Leghaemoglobin is found only in the nodules and is not produced by either the bacterium or the plant when grown alone.

In these leguminous associations the bacteria usually are Rhizobium species, but the root nodules of soybeans, chickpea and some other legumes are formed by small-celled rhizobia termed Bradyrhizobium. Nodules on some tropical leguminous plants are formed by yet other genera. In all cases the bacteria "invade" the plant and cause the formation of a nodule by inducing localised proliferation of the plant host cells. Yet the bacteria always remain separated from the host cytoplasm by being enclosed in a membrane - a necessary feature in symbioses.

Part of a crushed root nodule of a pea plant, showing four root cells containing colonies of Rhizobium. The nuclei (n) of two root cells are shown; cw indicates the cell wall that separates two plant cells. Although it cannot be seen clearly in this image, the bacteria occur in clusters which are enclosed in membranes, separating them from the cytoplasm of the plant cells.

In nodules where nitrogen-fixation is occurring, the plant tissues contain the oxygen-scavenging molecule, leghaemoglobin (serving the same function as the oxygen-carrying haemoglobin in blood). The function of this molecule in nodules is to reduce the amount of free oxygen, and thereby to protect the nitrogen-fixing enzyme nitrogenase, which is irreversibly inactivated by oxygen.

Some excellent images and discussion of these leguminous associations can be found at:

Frankia is a genus of the bacterial group termed actinomycetes - filamentous bacteria that are noted for their production of air-borne spores. Included in this group are the common soil-dwelling Streptomyces species which produce many of the antibiotics used in medicine.

Frankia species are slow-growing in culture, and require specialised media, suggesting that they are specialised symbionts. They form nitrogen-fixing root nodules (sometimes called actinorhizae) with several woody plants of different families, such as alder (Alnus species), sea buckthorn (Hippophae rhamnoides, which is common in sand-dune environments) and Casuarina (a Mediterranean tree genus).

Alder and the other woody hosts of Frankia are typical pioneer species that invade nutrient-poor soils. These plants probably benefit from the nitrogen-fixing association, while supplying the bacterial symbiont with photosynthetic products.

The photosynthetic cyanobacteria often live as free-living organisms in pioneer habitats such as desert soils or as symbionts with lichens in other pioneer habitats. They also form symbiotic associations with other organisms such as the water fern Azolla, and cycads.The association with Azolla, where cyanobacteria (Anabaena azollae) are harboured in the leaves, has sometimes been shown to be important for nitrogen inputs in rice paddies, especially if the fern is allowed to grow and then ploughed into the soil to release nitrogen before the rice crop is sown. A symbiotic association of cyanobacteria with cycads is shown below. The first image shows a pot-grown plant. The second image shows a close-up of the soil surface in this pot. Short, club-shaped, branching roots have grown into the aerial environment. These aerial roots contain a nitrogen-fixing cyanobacterial symbiont.

In addition to these intimate and specialised symbiotic associations, there are several free-living nitrogen-fixing bacteria that grow in close association with plants. For example, Azospirillum species have been shown to fix nitrogen when growing in the root zone (rhizosphere) or tropical grasses, and even of maize plants in field conditions. Similarly, Azotobacter species can fix nitrogen in the rhizosphere of several plants. In both cases the bacteria grow at the expense of sugars and other nutrients that leak from the roots. However, these bacteria can make only a small contribution to the nitrogen nutrition of the plant, because nitrogen-fixation is an energy-expensive process, and large amounts of organic nutrients are not continuously available to microbes in the rhizosphere.

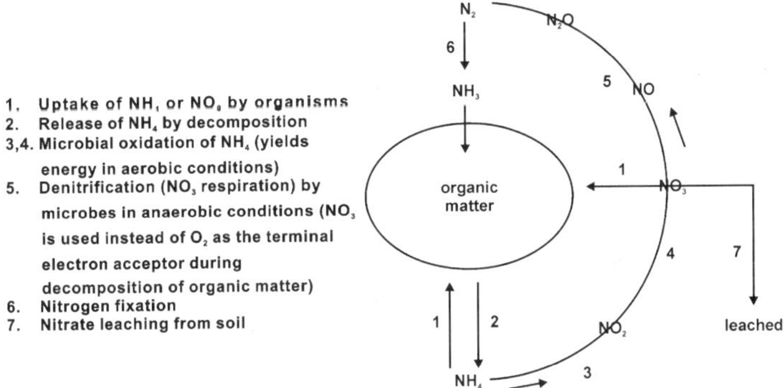

This limitation may not apply to the bacteria that live in root nodules or other intimate symbiotic associations with plants. It has been estimated that nitrogen fixation in the nodules of clover roots or other leguminous plants may consume as much as 20 per cent of the total photosynthate.

THE NITROGEN CYCLE

At any one time a large proportion of the total fixed nitrogen will be locked up in the biomass or in the dead remains of organisms. So, the only nitrogen available to support new growth will be that which is supplied by nitrogen fixation from the atmosphere or by the release of ammonium or simple organic nitrogen compounds through the decomposition of organic matter.

Nitrification

The term nitrification refers to the conversion of ammonium to nitrate (pathway 3-4). This is brought about by the nitrifying bacteria, which are specialised to gain their energy by oxidising ammonium, while using CO2 as their source of carbon to synthesise organic compounds. Organisms of this sort are termed chemoautotrophs - they gain their energy by chemical oxidations (chemo-) and they are autotrophs (self-feeders) because they do not depend on pre-formed organic matter. In principle the oxidation of ammonium by these bacteria is no different from the way in which humans gain energy by oxidising sugars. Their use of CO_2 to produce organic matter is no different in principle from the behaviour of plants.

The nitrifying bacteria are found in most soils and waters of moderate pH, but are not active in highly acidic soils. They almost always are found as mixed-species communities (termed consortia) because some of them - *e.g.* Nitrosomonas species - are specialised to convert ammonium to nitrite (NO_2-) while others - *e.g.* Nitrobacter species - convert nitrite to nitrate (NO_3-). In fact, the accumulation of nitrite inhibits Nitrosomonas, so it depends on Nitrobacter to convert this to nitrate, whereas Nitrobacter depends on Nitrosomonas to generate nitrite.

The nitrifying bacteria have some important environmental consequences, because they are so common that most of the ammonium in oxygenated soil or natural waters is readily converted to nitrate. Most plants and microorganisms can take up either nitratc or ammonium. However, process of nitrification has some undesirable consequences. The ammonium ion (NH_4+) has a positive charge and so is readily adsorbed onto the negatively charged clay colloids and soil organic matter, preventing it from being washed out of the soil by rainfall. In contrast, the negatively charged nitrate ion is not held on soil particles and so can be washed down the soil profile - the process termed leaching. In this way, valuable nitrogen can be lost from the soil, reducing the soil fertility. The nitrates can then accumulate in groundwater, and ultimately in drinking water. There are strict regulations governing the amount of nitrate that can be present in drinking water, because nitrates can be reduced to highly reactive nitrites by microorganisms in the anaerobic conditions of the gut. Nitrites are absorbed from the gut and bind to haemoglobin, reducing its oxygen-carrying capacity. In young babies this can lead to respiratory distress - the condition known as "blue baby syndrome". Nitrite in the gut also can react with amino compounds, forming highly carcinogenic nitrosamines.

Denitrification

Denitrification refers to the process in which nitrate is converted to gaseous compounds (nitric oxide, nitrous oxide and N_2) by microorganisms. The sequence usually involves the production of nitrite (NO_2-) as an intermediate step.Several types of bacteria perform this conversion when growing on organic matter in anaerobic conditions. Because of the lack of oxygen for normal aerobic respiration, they use nitrate in place of oxygen as the terminal electron acceptor. This is termed anaerobic respiration and can be illustrated as follows:

In aerobic respiration (as in humans), organic molecules are oxidised to obtain energy, while oxygen is reduced to water:

$$C_6H_{12}0_6 + 6\ 0_2 = 6\ CO_2 + 6\ H_2O + \text{energy}$$

In the absence of oxygen, any reducible substance such as nitrate (NO_3-) could serve the same role and be reduced to nitrite, nitric oxide, nitrous oxide or N_2. Thus, the conditions in which we find denitrifying organisms are characterised by

- A supply of oxidisable organic matter, and
- Absence of oxygen but availability of reducible nitrogen sources.

A mixture of gaseous nitrogen products is often produced because of the stepwise use of nitrate, nitrite, nitric oxide and nitrous oxide as electron acceptors in anaerobic respiration. The common denitrifying bacteria include several species of Pseudomonas, Alkaligenes and Bacillus. Their activities result in substantial losses of nitrogen into the atmosphere, roughly balancing the amount of nitrogen fixation that occurs each year.

PHOSPHORUS AND SULPHUR

PHOSPHORUS

Phosphorus, (from the Greek language *phôs* meaning "light", and *phoros* meaning "bearer"), is the chemical element in the periodic table that has the symbol P and atomic number 15. A multivalent non-metal of the nitrogen group, phosphorus is commonly found in inorganic phosphate rocks and in all living cells. Due to its high reactivity, it is never found as a free element in nature. It emits a faint glow upon exposure to oxygen, occurs in several allotropic forms, and is an essential element for living organisms. The most important commercial use of phosphorus is in the production of fertilizers. It is also widely used in explosives, nerve agents, friction matches, fireworks, pesticides, toothpaste, and detergents.

Phosphorus (P), the 15th element on the periodic table with an atomic weight of 30.974, is an essential nutrient for all life forms. Phosphorus plays a role in deoxyribonucleic acid (DNA), ribonucleic acid (RNA), adenosine diphosphate (ADP), and adenosine triphosphate (ATP). Phosphorus is required for these necessary components of life to occur.

Phosphate is usually not readily available for uptake in soils. Phosphate is only freely soluble in acid solutions and under reducing conditions. In the soil it is rapidly immobilized as calcium or iron phosphates. Most of the phosphorus in soils is adsorbed to soil particles or incorporated into organic matter.

Phosphorus in freshwater and marine systems exists in either a particulate phase or a dissolved phase. Particulate matter includes living and dead plankton, precipitates of phosphorus, phosphorus adsorbed to particulates, and amorphous phosphorus. The dissolved phase includes inorganic phosphorus (generally in the soluble orthophosphate form), organic phosphorus excreted by organisms, and macromolecular colloidal phosphorus.

The organic and inorganic particulate and soluble forms of phosphorus undergo continuous transformations. The dissolved phosphorus (usually as orthophosphate) is assimilated by phytoplankton and altered to organic phosphorus. The phytoplankton are then ingested by detritivores or zooplankton. Over half of the organic phosphorus taken up by zooplankton is excreted as inorganic P. Continuing the cycle, the inorganic P is rapidly assimilated by phytoplankton.

Lakes and reservoir sediments serve as phosphorus sinks. Phosphorus-containing particles settle to the substrate and are rapidly covered by sediment. Continuous accumulation of sediment will leave some phosphorus too deep within the substrate to be reintroduced to the water column. Thus, some phosphorus is removed permanently from biocirculation.

Health Effects

Phosphate: Phosphate itself does not have notable adverse health effects. However, phosphate levels greater than 1.0 may interfere with coagulation in water treatment plants. As a result, organic particles that harbour microorganisms may not be completely removed before distribution.

Environmental Effects

The growth of macrophytes and phytoplankton is stimulated principally by nutrients such as phosphorus and nitrogen. Nutrient-stimulated primary production is of most concern in lakes and estuaries, because primary production in flowing water is thought to be controlled by physical factors, such as light penetration, timing of flow, and type of substrate available, instead of by nutrients.

Freshwater system impacts: Generally, phosphorus (as orthophosphate) is the limiting nutrient in freshwater aquatic systems. That is, if all phosphorus is used, plant growth will cease, no matter how much nitrogen is available. The natural background levels of total phosphorus are generally less than 0.03 mg/l. The natural levels of orthophosphate usually range from 0.005 to 0.05 mg/l Many bodies of freshwater are currently experiencing influxes of phosphorus and nitrogen from outside sources.

The increasing concentration of available phosphorus allows plants to assimilate more nitrogen before the phosphorus is depleted. Thus, if sufficient phosphorus is available, elevated concentrations of nitrates will lead to algal blooms.

Although levels of 0.08 to 0.10 mg/l orthophosphate may trigger periodic blooms, long-term eutrophication will usually be prevented if total phosphorus levels and orthophosphate levels are below 0.5 mg/l and 0.05 mg/l, respectively.

Estuarine system impacts: In contrast to freshwater, nitrogen is generally the primary limiting nutrient in the seaward portions of estuarine systems. Here, nitrogen levels control the rate of primary production. If the system is supplied with high levels of nitrogen, algal blooms will occur. Systems may be phosphorus limited, however, or become so when nitrogen concentrations are high and N:P>16:1.

Freshwater and estuarine systems: Nutrient-induced production of aquatic plants in both freshwater and estuaries has several detrimental consequences:

Extensive growth of rooted aquatic macrophytes will interfere with navigation, aeration, and channel capacity.

Dead macrophytes and phytoplankton settle to the bottom of a water body, stimulating microbial breakdown processes that require oxygen. Eventually, oxygen will be depleted.

Sources

Non-Point Sources

Natural: Phosphate deposits and phosphate-rich rocks release phosphorus during weathering, erosion, and leaching. Phosphorus may be released from lake and reservoir bottom sediments during seasonal overturns.

Anthropogenic: The primary anthropogenic non-point sources of phosphorus include run-off from:

- Land areas being mined for phosphate deposits,
- Agricultural areas, and
- Urban/residential areas.

Because phosphorus has a strong affinity for soil, little dissolved phosphorus will be transported in run-off. Instead, the eroded sediments from mining and agricultural areas carry the adsorbed phosphorus to the water body. An additional source is the overboard discharge of phosphorus-containing sewage by boats.

Point sources

Sewage treatment plants provide most of the available phosphorus to surface water bodies. A normal adult excretes 1.3 - 1.5 g of phosphorus per day. Additional phosphorus originates from the use of industrial products, such as toothpaste, detergents, pharmaceuticals, and food-treating compounds. Primary treatment removes only 10 per cent of the phosphorus in the waste stream; secondary treatment removes only 30 per cent.

The remainder is discharged to the water body (Smith, 1990). Tertiary treatment is required to remove additional phosphorus from the water. The amount of additional phosphorus that can be removed varies with the success of the treatment technologies used. Available technologies include biological removal and chemical precipitation.

Mode of Transport

Phosphates are primarily discharged directly into the water body by sewage treatment plants. Phosphorus that is adsorped to sediment particles may be transported in overland flow.

Analytical Techniques

Total Phosphorus and Orthophosphate: Analysis involves two procedural steps:

Step 1: Digestion Methods

- *Perchloric Acid Digestion*: Recommended only for extremely difficult-to-analyze samples, such as sediments.
- *Nitric Acid-Sulphuric Acid Method:* Recommended for most samples.
- *Persulfate Oxidation Method:* This simple method should be cross-checked with one or more thorough techniques and adopted if results are identical.

Step 2: Colorimetric Methods

- *Ascorbic Acid Method*: Ammonium molybdate and potassium antimonyl tartrate react with orthophosphate to form a heteropoly acid that is reduced to molybdenum blue by ascorbic acid.
  - *Detection limits:* Ranges change with light path used.

Range (mg/l P)	Path (cm)
0.3 - 2.0	0.5
0.15 - 1.3	1.0
0.01 - 0.25	5.0

  - *Interferences:* Arsenates react with the molybdate to form a similar blue colour. Nitrite and hexavalent chromium interfere to yield results 3 per cent less than actual at 1 mg/l and 10 per cent to 15 per cent less than actual at 10 mg/l.
- *Automated Ascorbic Acid Reduction Method*: Ammonium molybdate and potassium antimonyl tartrate react with orthophosphate in an acid medium to form an antimony- phosphomolybdate complex that forms a blue colour suitable for photometric measurements when reduced by ascorbic acid.
  - *Detection limits:* 0.001 to 10.0 mg/l P when photometric measurements are performed at 650 to 600 in a 15mm tubular flow cell, or 880 nm in a 50mm tubular flow cell.
  - *Interferences:* >50 mg/l Fe(3+), 10 mg/l Cu, and 10 mg/l SiO2. Turbidity, colour may interfere. Arsenate provides a positive interference.

Notable Characteristics

Common phosphorus forms a waxy white solid that has a characteristic, disagreeable smell similar to that of garlic. Pure forms of the element are colorless and transparent. This non-metal is not soluble in water, but *is* soluble in carbon disulfide. Pure phosphorus ignites spontaneously in air, burning to produce phosphorus pentoxide.

Forms

Phosphorus is known to exist in three allotropic forms: white, red, and black (other allotropic forms are theoretically possible, but are not known). The most common are red and white phosphorus, both of which consist of networks of tetrahedrally arranged groups of four phosphorus atoms. The tetrahedra of white phosphorus form separate groups; the tetrahedra of red phosphorus are linked into chains. White phosphorus burns on contact with air and on exposure to heat or light.

Phosphorus also exists in kinetically and thermodynamically favored forms. They are separated by a transition temperature of -3.8 °C. One is known as the "alpha" form, the other "beta". Red phosphorus is comparatively stable and sublimates at a vapor pressure of 1 atm at 170 °C but burns from impact or frictional heating. A black phosphorus allotrope exists which has a structure similar to graphite – the atoms are arranged in hexagonal sheet layers and will conduct electricity.

Glow

The glow from phosphorus was the attraction of its discovery around 1669, but the mechanism for that glow was not fully described until 1974.

In 1974 the glow was explained by R. J. van Zee and A. U. Khan. A reaction with oxygen takes place at the surface of the solid (or liquid) phosphorus, forming short-lived molecules HPO and P_2O_2 and they both emit visible light. The reaction is slow and only very little of the intermediates is required to produce the luminescence, hence the extended time the glow continues in a stoppered jar.

Although the term phosphorescence is derived from phosphorus, the reaction is properly called luminescence (glowing by its own reaction, in this case chemoluminescence), not phosphorescence (re-emitting light that previously fell on it).

Applications

Concentrated phosphoric acids, which can consist of 70 per cent to 75 per cent P_2O_5 are very important to agriculture and farm production in the form of fertilizers. Global demand for fertilizers led to large increases in phosphate (PO_4^{3-}) production in the second half of the 20th century. Other uses;

- Phosphates are utilized in the making of special glasses that are used for sodium lamps.
- Phosphorus is widely used to make organophosphorus compounds, through the intermediates phosphorus chlorides and the two phosphorus sulfides: phosphorus pentasulfide, and phosphorus sesquisulfide. Organophosphorus compounds have many applications, including in plasticizers, flame retardants, pesticides, extraction agents, and water treatment.
- Phosphorus sesquisulfide is used in heads of strike-anywhere matches.
- This element is also an important component in steel production, in the making of phosphor bronze, and in many other related products.
- White phosphorus is used in military applications as incendiary bombs, for smoke-screening as smoke pots and smoke bombs, and in tracer ammunition.
- Red phosphorus is essential for manufacturing matchbook strikers, flares, and, most notoriously, methamphetamine.
- In trace amounts, phosphorus is used as a dopant for N-type semiconductors.
- ^{32}P and ^{33}P are used as radioactive tracers in biochemical laboratories.

Biological Role

Phosphorus is a key element in all known forms of life. Inorganic phosphorus in the form of the phosphate PO_4^{3-} plays a major role in biological molecules such as DNA and RNA where it forms part of the structural backbone of these molecules.

Living cells also utilize phosphate to transport cellular energy via adenosine triphosphate (ATP). Nearly every cellular process that uses energy gets it in the form of ATP. ATP is also important for phosphorylation, a key regulatory event in cells. Phospholipids are the main structural components of all cellular membranes.

Calcium phosphate salts are used by animals to stiffen their bones. An average person contains a little less than 1 kg of phosphorus, about three quarters of which is present in bones and teeth in the form of apatite. A well-fed adult in the industrialized world consumes and excretes about 1-3 g of phosphorus per day in the form of phosphate.

Phosphorus is an essential mineral macronutrient, which is studied extensively in soil conservation in order to understand plant uptake from soil systems.

In ecological terms, phosphorus is often a limiting nutrient in many environments, *i.e.* the availability of phosphorus governs the rate of growth of many organisms. In ecosystems an excess of phosphorus can be problematic, especially in aquatic systems.

History

Phosphorus (Greek *phosphoros* was the ancient name for the planet Venus) was discovered by German alchemist Hennig Brand in 1669 through a preparation from urine. Working in Hamburg, Brand attempted to distill salts by evaporating urine, and in the process produced a white material that glowed in the dark and burned brilliantly. Since that time, phosphorescence has been used to describe substances that shine in the dark without burning.

Phosphorus was first made commercially, for the match industry, in the 19th century, by distilling off phosphorus vapour from precipitated phosphates heated in a retort. The precipitated phosphates made from ground-up bones, that had been de-greased and treated with strong acids. This process became obsolete in the late 1890s when the Electric arc furnace was adapted to reduce phosphate rock.

Early matches used white phosphorus in their composition, which was dangerous due to its toxicity. Murders, suicides and accidental poisonings resulted from its use.

The electric furnace method allowed production to increase to the point phosphorus could be used in weapons of war. In World WarI it was used in incendiaries, smoke screens and tracer bullets.

Today phosphorus production is larger than ever, used as a precursor for various chemicals, in particular the pesticide glyphosate sold under the brand name Roundup. Production takes place at large facilities and is transported heated to liquid form.

Occurrence

Due to its reactivity to air and many other oxygen containing substances, phosphorus is not found free in nature but it is widely distributed in many different minerals. Phosphate rock, which is partially made of apatite (an impure tri-calcium phosphate mineral), is an important commercial source of this element. Large deposits of apatite are located in China, Russia, Morocco, Florida, Idaho, Tennessee, Utah, and elsewhere.

There are however concerns over how long these phosphorus deposits will last. The United States will deplete its deposits around 2035. China and Morocco have the largest known deposits today, but they too will eventually be depleted. During that depletion there could be a serious problem for the world's food production since phosphorus is such an essential ingredient in fertilizers.

The white allotrope can be produced using several different methods. In one process, tri-calcium phosphate, which is derived from phosphate rock, is heated in an electric or fuel-fired furnace in the presence of carbon and silica. Elemental phosphorus is then liberated as a vapor and can be collected under phosphoric acid.

Precautions

Organic compounds of phosphorus form a wide class of materials, some of which are extremely toxic. Fluorophosphate esters are among the most potent neurotoxins known. A wide range of organophosphorus compounds are used for their toxicity to certain organisms as pesticides (herbicides, insecticides, fungicides etc) and weaponized as nerve agents. Most inorganic phosphates are relatively non-toxic and essential nutrients.

The allotrope white phosphorus should be kept under water at all times as it presents a significant fire hazard due to its extreme reactivity to atmospheric oxygen, and it should only be manipulated with forceps since contact with skin can cause severe burns. Chronic white phosphorus poisoning of unprotected workers leads to necrosis of the jaw called "phossy-jaw". Ingestion of white phosphorus may cause a medical condition known as "Smoking Stool Syndrome".

When the white form is exposed to sunlight or when it is heated in its own vapor to 250 °C, it is transmuted to the red form, which does not phosphoresce in air. The red allotrope does not spontaneously ignite in air and is not as dangerous as the white form. Nevertheless, it should be handled with care because it does revert to white phosphorus in some temperature ranges and it also emits highly toxic fumes that consist of phosphorus oxides when it is heated.

Upon exposure to elemental phosphorus, in the past it was suggested to wash the affected area with 2 per cent copper sulfate solution to form harmless compounds that can be washed away. The manual suggests instead "a bicarbonate solution to neutralize phosphoric acid, which will then allow removal of visible WP. Particles often can be located by their emission of smoke when air strikes them, or by their phosphorescence in the dark. In dark surroundings, fragments are seen as luminescent spots." Then, "Promptly debride the burn if the patient's condition will permit removal of bits of WP which might be absorbed later and possibly produce systemic poisoning. DO NOT apply oily-based ointments until it is certain that all WP has been removed. Following complete removal of the particles, treat the lesions as thermal burns." As white phosphorus readily mixes with oils, any oily substances or ointments are disrecommended until the area is thoroughly cleaned and all white phosphorus removed.

Isotopes

Radioactive isotopes of phosphorus include:

^{32}P; a beta-emitter (1.71 MeV) with a half-life of 14.3 days which is used routinely in life-science laboratories, primarily to produce radiolabeled DNA and RNA probes, *e.g.* for use in Northern blots or Southern blots. Because the high energy beta particles produced penetrate skin and corneas, and because any ^{32}P ingested, inhaled, or absorbed is readily incorporated into bone and nucleic acids, OSHA requires that a lab coat, disposable gloves, and safety glasses or goggles be worn when working with ^{32}P, and that working directly over an open container be avoided in order to protect the eyes.

Monitoring personal, clothing, and surface contamination is also required. In addition, due to the high energy of the beta particles, shielding this radiation with the normally used dense materials (*e.g.* lead), gives rise to secondary emission of X-rays via a process known as Bremsstrahlung, meaning braking radiation. Therefore shielding must be accomplished with low density materials, *e.g.* Plexiglas, acrylic, Lucite, plastic, wood, or water.

^{33}P; a beta-emitter (0.25 MeV) with a half-life of 25.4 days. It is used in life-science laboratories in applications in which lower energy beta emissions are advantageous such as DNA sequencing.

Compounds

- Ammonium phosphate ($(NH_4)_3PO_4$)
- Phosphine (Phosphorus Trihydride PH_3)
- Calcium phosphate ($Ca_3(PO_4)_2$)
- Calcium dihydrogen phosphate ($Ca(H_2PO_4)_2$)
- Iron(III) phosphate ($FePO_4$)
- Iron(II) phosphate ($Fe_3(PO_4)_2$)
- Hypophosphorous acid (H_3PO_2)

NATURE OF PHOSPHORUS IN SOILS

Phosphorus (P) is an essential element classified as a macronutrient because of the relatively large amounts of P required by plants. Phosphorus is one of the three nutrients generally added to soils in fertilizers. One of the main roles of P in living organisms is in the transfer of energy. Organic compounds that contain P are used to transfer energy from one reaction to drive another reaction within cells.

Adequate P availability for plants stimulates early plant growth and hastens maturity. Although P is essential for plant growth, mismanagement of soil P can pose a threat to water quality. The concentration of P is usually sufficiently low in fresh water so that algae growth is limited.

When lakes and rivers are polluted with P, excessive growth of algae often results. High levels of algae reduce water clarity and can lead to decreases in available dissolved oxygen as the algae decays, conditions that can be very detrimental to game fish populations.

The Phosphorus Cycle

Phosphate is taken up by plants from soils, utilized by animals that consume plants, and returned to soils as organic residues decay in soils. Much of the phosphate used by living organisms becomes incorporated into organic compounds.

When plant materials are returned to the soil, this organic phosphate will slowly be released as inorganic phosphate or be incorporated into more stable organic materials and become part of the soil organic matter. The release of inorganic phosphate from organic phosphates is called mineralization and is caused by microorganisms breaking down organic compounds. The activity of microorganisms is highly influenced by soil temperature and soil moisture. The process is most rapid when soils are warm and moist but well drained. Phosphate can potentially be lost through soil erosion and to a lesser extent to water running over or through the soil.

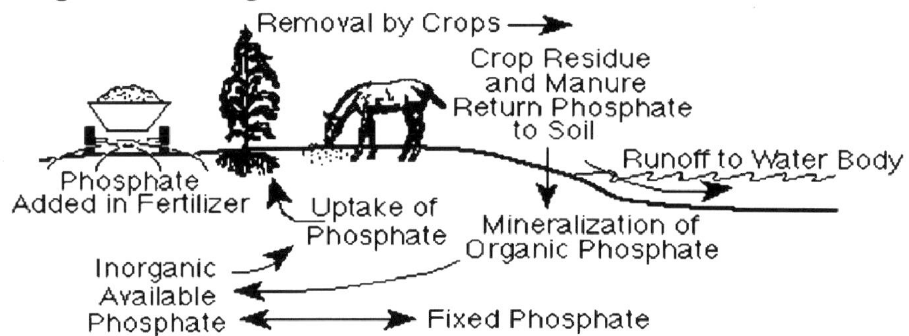

Many phosphate compounds are not very soluble in water; therefore, most of the phosphate in natural systems exists in solid form. However, soil water and surface water (rivers and lakes) usually contain relatively low concentrations of dissolved (or soluble) phosphorus.

Depending on the types of minerals in the area, bodies of water usually contain about 10 ppb or more of dissolved P as orthophosphate. Water bodies may also contain organic P and phosphate attached to small particles of sediment.

Total phosphorus in water is all of the phosphorus in solution regardless of its form and is often the form reported in water quality studies. Algal available or bioavailable phosphorus is P that is estimated to be available to organisms like algae that are present in a lake or river. This is usually estimated by a chemical test which is designed to measure the dissolved P and the particulate P that are easily available. This is a measure of the P that is of immediate concern to water quality.

Forms of Phosphorus in Soils

In soils P may exist in many different forms. In practical terms, however, P in soils can be thought of existing in 3 "pools":

- Solution P
- Active P
- Fixed P

The *solution P* pool is very small and will usually contain only a fraction of a pound of P per acre. The solution P will usually be in the orthophosphate form, but small amounts of organic P may exist as well. Plants will only take up P in the orthophosphate form. The solution P pool is important because it is the pool from which plants take up P and is the only pool that has any measurable mobility. Most of the P taken up by a crop during a growing season will probably have moved only an inch or less through the soil to the roots. A growing crop would quickly deplete the P in the soluble P pool if the pool was not being continuously replenished.

The *active P* pool is P in the solid phase which is relatively easily released to the soil solution, the water surrounding soil particles. As plants take up phosphate, the concentration of phosphate in solution is decreased and some phosphate from the active P pool is released. Because the solution P pool is very small, the active P pool is the main source of available P for crops. The ability of the active P pool to replenish the soil solution P pool in a soil is what makes a soil fertile with respect to phosphate. An acre of land may contain several pounds to a few hundred pounds of P in the active P pool. The active P pool will contain inorganic phosphate that is attached (or adsorbed) to small particles in the soil, phosphate that reacted with elements such as calcium or aluminum to form somewhat soluble solids, and organic P that is easily mineralized. Adsorbed phosphate ions are held on active sites on the surfaces of soil particles. The amount of phosphate adsorbed by soil increases as the amount of phosphate in solution increases and vice versa. Soil particles can act either as a source or a sink of phosphate to the surrounding water depending on conditions. Soil particles with low levels of adsorbed P that are eroded into a body of water with relatively high levels of dissolved phosphate may adsorb phosphate from the water, and vice versa.

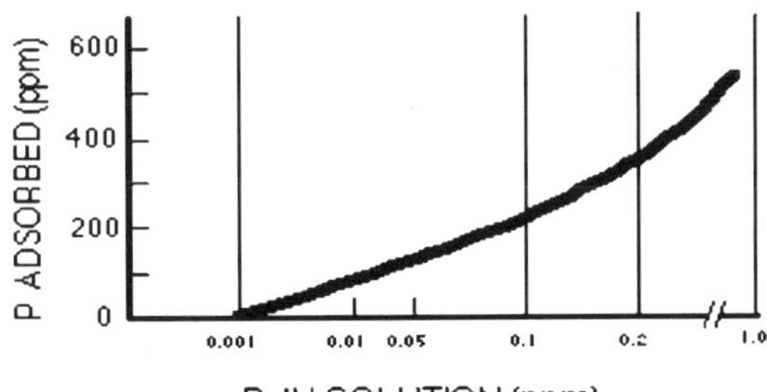

Fig. Relationship between P adsorbed by soil and P in solution.

The *fixed P* pool of phosphate will contain inorganic phosphate compounds that are very insoluble and organic compounds that are resistant to mineralization by microorganisms in the soil. Phosphate in this pool may remain in soils for years without being made available to plants and may have very little impact on the fertility of a soil. The inorganic phosphate compounds in this fixed P pool are more crystalline in their structure and less soluble than those compounds considered to be in the active P pool. Some slow conversion between the fixed P pool and the active P pool does occur in soils.

Fate of Phosphorus Added to Soils

The phosphate in fertilizers and manure is initially quite soluble and available. Most phosphate fertilizers have been manufactured by treating rock phosphate (the phosphate-bearing mineral that is mined) with acid to make it more soluble. Manure contains soluble phosphate, organic phosphate, and inorganic phosphate compounds that are quite available. When the fertilizer or manure phosphate comes in contact with the soil, various reactions begin occurring that make the phosphate less soluble and less available. The rates and products of these reactions are dependent on such soil conditions as pH, moisture content, temperature, and the minerals already present in the soil.

As a particle of fertilizer comes in contact with the soil, moisture from the soil will begin dissolving the particle. Dissolving of the fertilizer increases the soluble phosphate in the soil solution around the particle and allows the dissolved phosphate to move a short distance away from the fertilizer particle. Movement is slow but may be increased by rainfall or irrigation water flowing through the soil. As phosphate ions in solution slowly migrate away from the fertilizer particle, most of the phosphate will react with the minerals within the soil. Phosphate ions generally react by adsorbing to soil particles or by combining with elements in the soil such as calcium (Ca), magnesium (Mg), aluminum (Al), and iron (Fe), and forming compounds that are solids. The adsorbed phosphate and the newly formed solids are relatively available to meet crop needs.

Gradually reactions occur in which the adsorbedphosphate and the easily dissolved compounds of phosphate form more insoluble compounds that cause the phosphate to be become fixed and unavailable. Over time this results in a decrease in soil test P. The mechanisms for the changes in phosphate are complex and involve a variety of compounds. In alkaline soils (soil pH greater than 7) Ca is the dominant cation (positive ion) that will react with phosphate. A general sequence of reactions in alkaline soils is the formation of dibasic calcium phosphate dihydrate, octocalcium phosphate, and hydroxyapatite. The formation of each product results in a decrease in solubility and availability of phosphate. In acidic soils (especially with soil pH less than 5.5) Al is the dominant ion that will react with phosphate. In these soils the first products formed would be amorphous Al and Fe phosphates, as well as some Ca phosphates. The amorphous Al and Fe phosphates gradually change into compounds that resemble crystalline variscite (an Al phosphate) and strengite (an Fe phosphate). Each of these reactions will result in very insoluble compounds of phosphate that are generally not available to plants. Reactions that reduce P availability occur in all ranges of soil pH but can be very pronounced in alkaline soils ($pH > 7.3$) and in acidic soils ($pH < 5.5$). Maintaining soil pH between 6 and 7 will generally result in the most efficient use of phosphate.

Adding to the active P pool through fertilization will also increase the amount of fixed P. Depleting the active pool through crop uptake may cause some of the fixed P to slowly become active P. The conversion of available P to fixed P is partially the reason for the low efficiency of P fertilizers. Most of the P fertilizer applied to the soil will not be utilized by the crop in the first season. Continued application of more P than the crops utilize increases the fertility of the soil, but much of the added P becomes fixed and unavailable.

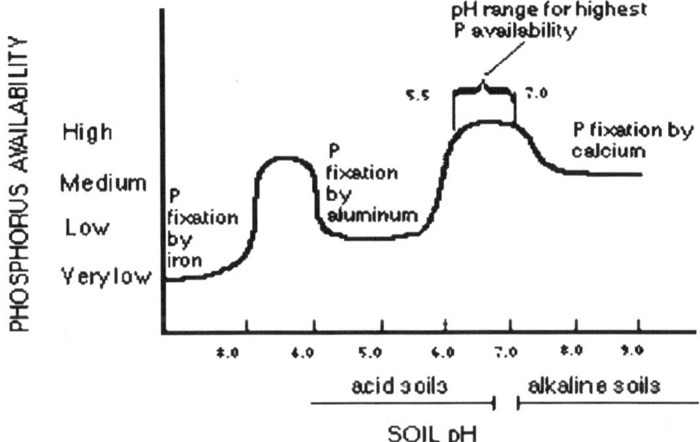

Fig. The availability of phosphorus is affected by soil pH

Most fine-to medium-textured soils have large capacities to hold phosphate by adsorption and precipitation. Occasionally the question of how much phosphate a soil can hold is asked, especially when high loading rates of P are expected or have occurred. Soils differ in their phosphate holding capacity. Fine-textured soils can generally hold hundreds of pounds of phosphate per acre. while coarse-textured soils can generally hold much less phosphate due to the more inert character of sand particles as compared to clay particles. In addition the subsoil of many soils often has an even greater capacity to hold phosphate than does the corresponding surface soil. However, an important aspect of the ability of a soil to hold phosphate is that a soil cannot hold increasing amounts of phosphate in the solid phase without also increasing soil solution phosphate. Increased amounts of phosphate in solution will potentially cause more phosphate to be lost to water running over the soil surface or leaching through the soil. Loading soils with very high levels of phosphate will generally not hurt crops but may result in increased phosphate movement to nearby bodies of water.

Predicting the Availability of Phosphorus in Soils

Soils may contain several hundred to several thousand pounds of phosphate per acre. However, much of the phosphate in soils is not available to growing plants. Phosphate in the soil solution P pool is immediately available but the amount is very small in comparison to the total P in soils. The active P pool is phosphorus that can be released into solution but is generally small in comparison to the fixed P.

To determine the need for supplemental P, soil tests are often used to estimate how much phosphate will be available for a crop. The most common way of determining P availability is to mix a small amount of soil with an extracting solution that contains an acid and/or complexing agent that will remove some of the phosphate from the soil particles. The extracting solution and soil are separated by filtration and the amount of P extracted is determined. In alkaline soils, a basic solution may be used as the extractant because an acidic solution will be neutralized by the alkaline soil and be less effective in extracting P. Calibration studies have been done to correlate crop response to fertilizer additions in soils with various soil test levels of P. Using the calibration data, recommendations can be made as to the amounts of phosphate fertilizer that will most likely give optimum yields.

Soil Phosphorus and Water Quality

Phosphorus is a somewhat unique pollutant in that it is an essential element, has low solubility, and is not toxic itself, but may have detrimental effects on water quality at quite low concentrations. There is considerable concern about P being lost from soils and transported to nearby streams and lakes. Several chemical properties of soil P have important implications for the potential loss of P to surface water.

Phosphorus in soils is almost entirely associated with soil particles. When soil particles are carried to a river or lake, P will be contained in this sediment. When the sediment reaches a body of water it may act as a sink or a source of P in solution. In either case, it is a potential source of P that may eventually be released.

Most soils have a large capacity to retain P. Even large additions of P will be mostly retained by soils provided there is adequate contact with the soil.

Increasing the amounts of phosphate in soils results in increased levels of phosphate in soil solutions. This will generally result in small but potentially important increases in the amounts of phosphate in water that passes over or through soils.

Phosphate in soils is associated more with fine particles than coarse particles. When soil erosion occurs, more fine particles are removed than coarse particles, causing sediment leaving a soil through erosion to be enriched in P.

PHOSPHORUS CONCERNS IN THE ENVIRONMENT

Eutrophication can be caused by the nutrient enrichment of a water body. Nutrient movement in run-off and erosion from agricultural non-point sources is a resource management concern. The movement of phosphorus in run-off from agricultural land to surface water can accelerate eutrophication. Undesirable aquatic plant growth results from additions of phosphorus to the water. The net result of the eutrophic condition and excess plant growth in water is the depletion of oxygen in the water due to the heavy oxygen demand by microorganisms as they decompose the organic material. Little attention has been given to management strategies to minimize the non-point movement of P in the landscape because of the easier identification and control of point source inputs of P to surface waters and a lack of direct human health risks associated with eutrophication. Phosphorus is generally the limiting nutrient in fresh water systems and any increase in P is usually results in more aquatic vegetation. Society is concerned about maintaining clean drinking water. This concern now includes a cost for removing the colour, taste, and odour associated with the high trophic condition and vegetation growth in surface water due to excess nutrients.

PHOSPHORUS MOVEMENT IN THE LANDSCAPE

Phosphorus movement in run-off occurs as particulate P and dissolved P. Particulate P is attached to mineral and organic sediment as it moves with the run-off. Dissolved P is in the water solution. In general, particulate P is the major portion (75-90 per cent) of the P transported in run-off from cultivated land. Dissolved P makes up a larger portion of the total P in run-off from non-cultivated lands such as pastures and fields with reduced tillage. In terms of their impact on eutrophication of water bodies, particulate P becomes less available to algae and plant uptake than dissolved P because of the chemical form it has with the mineral (particularly iron, aluminum, and calcium) and organic compounds. The availability of particulate P to plants and algae is variable, ranging from 10 to 90 per cent of the total P. yet can represent a long-term source of P for algae and plant uptake from the water body. Dissolved P is 100 per cent bioavailable to plants. Added together, the bioavailable portion of particulate P and the dissolved P represents the phosphorus that promotes eutrophication of surface waters.

The method by which P in both particulate and dissolved form moves within the landscape are simplified in the following description. Eroding soil material is transported by run-off. During detachment and movement of sediment in run-off, the finer clay-sized fraction of the source material are preferentially eroded. The P content and reactivity of the eroded material to P are usually greater than the source soil from which it was eroded. The suspended sediment in the run-off can rapidly adsorb the dissolved P in the run-off water.

As run-off moves from the landscape and towards the water body there is generally a progressive dilution of P through additions of water and a reduction in the amount of sediment carried because of sediment deposition. Phosphorus may become more bioavailable by the sorption and Resorption processes and by the preferential transport of claysized material as sediment moves over the landscape.

The movement of dissolved P begins with the Resorption, dissolution, and extraction of P from the soil, plant, and organic material. These processes occur when rain and run-off water interact with the thin layer of surface soil (0.05 to 0.10 inches). Some water infiltrates into the soil and percolates through the profile where desorption of P will result in a low dissolved concentration in subsurface and return flow. High dissolved P concentration can be expected in the water percolating through organic, coarse-textured, and oxygen depleted (reduced), water-logged soils. Soil pH also affects the movement and availability of phosphorus.

The interaction between the particulate and dissolved P in the run-off is very dynamic and the mechanism of transport is complex. Therefore, it is difficult to predict the transformation and ultimate fate of P as it moves through the landscape.

THE CONCEPT

The purpose of the Phosphorus Index is to provide field staffs, watershed planners, and land users with a tool to assess the various landforms and management practices for potential risk of phosphorus movement to water bodies. The ranking of Phosphorus Index identifies sites where the risk of phosphorus movement may be relatively higher than that of other sites. When the parameters of the index are analyzed, it will become apparent that an individual parameter or parameters may be influencing the index disproportionately.

For the first version, the Phosphorus Index will be a 8 x 5 matrix using a limited number of landform site and management characteristics. The input to the matrix will be readily accessible field data. This index will be used as a tool for understanding the contribution that individual landform and management parameters have towards risk of phosphorus movement and will provide a method for developing management guidelines for phosphorus at the site to lessen their impact on water quality.

The P Index is a simple 8 by 5 matrix that relates site characteristics with a range of value categories. This first version (Table) of the P Index has eight site characteristics. The eight characteristics are:

- Soil erosion,
- Irrigation erosion,
- Run-off class,
- Soil P test,
- P fertilizer application rate,

- P fertilizer application method,
- Organic P source application rate,
- Organic P source application method.

Each site charactersticts is rated on the basis of:

- None,
- Low,
- Medium,
- High,
- Very high.

There are eight site characteristics used in the P Index to assess a particular site. Each site characteristic is rated None, Low, Medium, High, or Very High by determining the range for each category. Although ranges have been given in this version of the index, it is imperative that each user establish their own range of values for each characteristic. Each user group should consider the cropping, soil, management conditions in the area designated to use the Phosphorus Index. Consideration need to be made as to base conditions for soil erosion, soil phosphorus test levels, and crop requirements. From these local considerations the ranges of the values category can be made. Any user group should include NRCS, Extension Service, University and ARS researchers, as well as state natural resource agencies and other groups interested in water quality. The definition of each of the eight site characteristics are:

SOIL EROSION

Soil erosion is defined as the loss of soil along the slope or unsheltered distance caused by the processes of water and wind. Soil erosion is estimated from erosion prediction models currently used (USLE or RUSLE for water erosion and WEQ for wind erosion). Erosion induced by irrigation is calculated by other convenient methods. The value category is given in tons of soil loss per acre per year (ton/acre/year). These soil loss prediction models do not predict sediment transport and delivery to a water body. The prediction models are used in this index to indicate a movement of soil, thus potential for sediment and attached phosphorus movement across the slope or unsheltered distance and towards a water body.

IRRIGATION EROSION

Potential P loss resulting from furrow irrigation induced erosion is considered by inclusion of a rating system based on soil susceptibility to particle detachment by hydraulic shear and flow rate of water in the furrow. The susceptibility to detachment is given by a relative ranking of soil erodibility classes under furrow irrigation (table). These classes are an initial attempt at a relative ranking based on inherent stable and static soil properties (*i.e.*, texture and clay mineralogy). There are temporal variations in the relative erodibility and actual amount of erosion with furrow erosion. These changes in erodibility are a function in the soil properties or a result of management. However, no attempt is made to consider temporal soil properties nor management factors in the rating. The introduced flow rate in the furrow (Q) is given by the irrigation plan and recorded as gallons per minute (gal/min). The furrow slope (S) of the site is given as a percentage (feet per 100 feet). The product of flow rate (Q) and slope (S) is used to determine the value category.

RUN-OFF CLASS

The run-off class of the site can be determined from soil survey data. Guidance in determining the run-off class is based on the soil saturated hydraulic conductivity (K_{Sat}) and the per cent slope of the site. Another method to determine run-off class is using the NRCS curve number (CN) method. The major factors that determine curve number are the hydrologic soil group. Landscape cover type, conservation treatment, hydrologic condition, and antecedent run-off conditions.

SOIL P TEST

A soil sample from the site is necessary to assess the level of "available P" in the surface layer of the soil. The available P is the level customarily given in a sail test analysis by the Cooperative Extension Service or commercial soil test laboratories.

The user of the P Index must determine the ranges of soil test P in each value category (Low, Medium, High, and Excessive). These ranges of soil test P values will vary by soil test method and region. The soil test level for "available P" does not ascertain the total P in the surface soil. It does however, give an indication of the amount of total P that may be present because of the general relationship between the forms of P (organic, adsorbed, and labile P) and the solution P available for crop uptake.

P FERTILIZER APPLICATION RATE

The P fertilizer application rate is the amount, in pounds per acre (lb/acre), of phosphate fertilizer (P_20_5) that is applied to the soil. This phosphate fertilizer does not include phosphorus from organic sources.

P FERTILIZER APPLICATION METHOD

The manner in which P fertilizer is applied to the soil and the amount of time that the fertilizer is exposed on the soil surface until crop utilization effects potential P movement. Incorporation implies that the fertilizer P is buried below the soil surface at a minimum of two inches.

The value categories of increasing severity, Low to Very High, depict the longer surface exposure time between fertilizer application, incorporation, and crop utilization.

ORGANIC P SOURCE APPLICATION RATE

The organic P application rate is the amount, in pounds per acre (lb/acre), of potential phosphate (P_2O_5) that is contained in the manure and applied to the soil. An analysis of the organic material is necessary to determine the potential phosphate content of the manure. The P content by analysis is generally considered to be completely plant available. This organic phosphate source does not include phosphorus from fertilizer sources.

ORGANIC P SOURCE APPLICATION METHOD

The manner in which organic P material is applied to the soil and the time that the organic material is exposed on the soil surface until crop utilization can determine potential P movement. Incorporation implies that the organic P material is buried below the soil surface at a minimum of two inches. The value categories of increasing severity, Low to Very High, depict the longer surface exposure time between organic P material application, incorporation, and crop utilization.

SULPHUR

Sulphur or sulphur is the chemical element in the periodic table that has the symbol S and atomic number 16. It is an abundant, tasteless, odorless, multivalent non-metal. Sulphur, in its native form, is a yellow crystaline solid. In nature, it can be found as the pure element or as sulfide and sulfate minerals. It is an essential element for life and is found in two amino acids. Its commercial uses are primarily in fertilizers but it is also widely used in gunpowder, matches, insecticides and fungicides.

NOTABLE CHARACTERISTICS

A piece of sulphur melts to a blood-red liquid. When burned, it emits a blue flame. At room temperature, sulphur is a soft bright yellow solid. Although sulphur is infamous for its smell—frequently compared to rotten eggs—the odour is actually characteristic of hydrogen sulfide (H_2S); elemental sulphur is odorless.

It burns with a blue flame that emits sulphur dioxide, notable for its peculiar suffocating odour. Sulphur is insoluble in water but soluble in carbon disulfide and to a lesser extent in other organic solvents such as benzene. Common oxidation states of sulphur include “2, +2, +4 and +6. Sulphur forms stable compounds with all elements except the noble gases.

APPLICATIONS

Sulphuric acid production is the major end use for sulphur, and consumption of sulphuric acid has been regarded as one of the best indices of a nation's industrial development. Sulphur is also used in batteries, detergents, the vulcanization of rubber, fungicides, and in the manufacture of phosphate fertilizers. Sulfites are used to bleach paper and as a preservative in wine and dried fruit. Because of its flammable nature, sulphur also finds use in matches, gunpowder, and fireworks. Sodium or ammonium thiosulfate are used as photographic fixing agents. Magnesium sulfate, better known as Epsom salts, can be used as a laxative, a bath additive, an exfoliant, or a magnesium supplement for plants. Sulphur is used as the light-generating medium in the rare lighting fixtures known as sulphur lamps.

BIOLOGICAL ROLE

The amino acids cysteine and methionine contain sulphur, as do all polypeptides, proteins, and enzymes which contain these amino acids. This makes sulphur a necessary component of all living cells. Disulfide bonds between polypeptides are very important in protein assembly and structure. Homocysteine and taurine are also sulphur containing amino acids but are not coded for by DNA nor are they part of the primary structure of proteins. Some forms of bacteria use hydrogen sulfide (H_2S) in the place of water as the electron donor in a primitive photosynthesis-like process. Sulphur is absorbed by plants via the roots from soil as the sulfate ion and reduced to sulfide before it is incorporated into cysteine and other organic sulphur compounds (sulphur assimilation). Inorganic sulphur forms a part of iron-sulphur clusters, and sulphur is the bridging ligand in the Cu_A site of cytochrome c oxidase. Sulphur is an important component of coenzyme A.

ENVIRONMENTAL IMPACT

The burning of coal and petroleum by industry and power plants liberates huge amounts of sulphur dioxide (SO_2) which reacts with atmospheric water and oxygen to produce sulphuric acid. This sulphuric acid is a component of acid rain, which lowers the pH of soil and freshwater bodies, resulting in substantial damage to the natural environment and chemical weathering of statues and architecture. Fuel standards increasingly require sulphur to be extracted from fossil fuels to prevent the formation of acid rain. This extracted sulphur is then refined and represents a large portion of sulphur production.

HISTORY

Homer mentioned "pest-averting sulphur" in the 8th century BC and in 424 BC, the tribe of Boeotia destroyed the walls of a city by burning a mixture of coal, sulphur, and tar under them. Sometime in the 12th century, the Chinese invented gun powder which is a mixture of potassium nitrate (KNO_3), carbon, and sulphur. Early alchemists gave sulphur its own alchemical symbol which was a triangle at the top of a cross. In the late 1770s, Antoine Lavoisier helped convince the scientific community that sulphur was an element and not a compound. In 1867, sulphur was discovered in underground deposits in Louisiana and Texas. The overlying layer of earth was quicksand, prohibiting ordinary mining operations. Therefore the Frasch process was utilized.

OCCURRENCE

Elemental sulphur can be found near hot springs and volcanic regions in many parts of the world, especially along the Pacific Ring of Fire. These occurrences are the basis for the traditional name brimstone, since sulphur could be found near the brims of volcanic craters. Such volcanic deposits are currently exploited in Indonesia, Chile, and Japan.

Significant desposits of elemental sulphur also exist in salt domes along the coast of the Gulf of Mexico, and in evaporites in eastern Europe and western Asia. The sulphur in these deposits is believed to come from the action of anaerobic bacteria on sulfate minerals, especially gypsum. Such deposits are the basis for commercial production in the United States, Poland, Russia, Turkmenistan, and Ukraine.

Sulphur extracted from oil, gas and the Athabasca Oil Sands has become a glut on the market, with huge stockpiles of sulphur in existence throughout Alberta.

COMPOUNDS

Many of the unpleasant odors of organic matter are based on sulphur-containing compounds such as methyl and ethyl mercaptan used to scent natural gas so that leaks are easily detectable. The odour of garlic and "skunk stink" are also caused by sulphur-containing organic compounds. However, not all organic sulphur compounds smell unpleasant; for example, grapefruit mercaptan, a sulphur-containing monoterpenoid is responsible for the characteristic scent of grapefruit.

Polymeric sulphur nitride has metallic properties even though it does not contain any metal atoms. This compound also has unusual electrical and optical properties. This polymer can be made from tetrasulphur tetranitride S_4N_4.

Inorganic Sulphur Compounds

- Sulfides (S^{2-}), a complex family of compounds usually derived from S^{2-}. Cadmium sulfide (CdS) is an example.
- Sulfates (SO_4^{2-}), the salts of sulphuric acid. Sulphuric acid also reacts with SO_3 in equimolar ratios to form pyrosulphuric acid ($H_2S_2O_7$).
- Sodium dithionite, $Na_2S_2O_4$, is the highly reducing dianion derived from hyposulphurous/dithionous acid.
- Sodium dithionate ($Na_2S_2O_6$).
- Polythionic acids ($H_2S_nO_6$), where *n* can range from 3 to 80.
- Sodium polysulfides (Na_2S_x)

- Sulphur hexafluoride, SF_6, a dense gas at ambient conditions, is used as non-reactive and non-toxic propellant
- Sulphur nitrides are chain and cyclic compounds containing only S and N. Tetrasulphur tetranitride S_4N_4 is an example.

ISOTOPES

Sulphur has 18 isotopes, of which four are stable: ^{32}S (95.02 per cent), ^{33}S (0.75 per cent), ^{34}S (4.21 per cent), and ^{36}S (0.02 per cent). Other than ^{35}S, the radioactive isotopes of sulphur are all short lived. ^{35}S is formed from cosmic ray spallation of ^{40}Ar in the atmosphere. It has a half-life of 87 days.

When sulfide minerals are precipitated, isotopic equilibration among solids and liquid may cause small differences in the äS-34 values of co-genetic minerals. The differences between minerals can be used to estimate the temperature of equilibration. The äC-13 and äS-34 of coexisting carbonates and sulfides can be used to determine the pH and oxygen fugacity of the ore-bearing fluid during ore formation.

In most forest ecosystems, sulfate is derived mostly from the atmosphere; weathering of ore minerals and evaporites also contribute some sulphur. Sulphur with a distinctive isotopic composition has been used to identify pollution sources, and enriched sulphur has been added as a tracer in hydrologic studies. Differences in the natural abundances can also be used in systems where there is sufficient variation in the ^{34}S of ecosystem components. Rocky Mountain lakes thought to be dominated by atmospheric sources of sulfate have been found to have different äS-34 values from lakes believed to be dominated by watershed sources of sulfate.

PRECAUTIONS

Carbon disulfide, carbon oxysulfide, hydrogen sulfide, and sulphur dioxide should all be handled with care.

Although sulphur dioxide is sufficiently safe to be used as a food additive in small amounts, at high concentrations it reacts with moisture to form sulphurous acid which in sufficient quantities may harm the lungs, eyes or other tissues. In creatures without lungs such as insects or plants, it otherwise prevents respiration.

Hydrogen sulfide is quite toxic (more toxic than cyanide). Although very pungent at first, it quickly deadens the sense of smell, so potential victims may be unaware of its presence until it is too late.

EFFECT OF SULPHUR ON WHEAT PLANT GROWTH

Sulphur is a building block of protein and a key ingredient in the formation of chlorophyll. Without adequate sulphur, crops can not possibly reach their full potential in terms of yield or protein content. Nor can they make efficient use of nitrogen, phosphorus and other vital elements.

Sulphur Content and Requirement of Wheat Plants

Wheat plants have a lesser requirement for sulphur than oilseed crops such as canola. A 2690 kg/ha (40 bu/ac) wheat crop will contain in the seed and straw 12 kg/ha (10 lb/ac) of sulphur, of which only 40 per cent is in the seed.

Sulphur (S)			
Yield	**Crop Part**	**Kg/ha**	**(lb/ac)**
Wheat	Seed	5	4
2690 Kg/ha	Straw	7	6
(40 Bu/ac)	Total	12	10
Canola	Seed	13	12
1960 Kg/ha	Straw	10	9

Time of Application

If sulphate fertilizers are applied in the fall, a portion of the sulphate could leach downward as much as 50 cm (20 in), depending on the amount of fall and spring precipitation and the texture of the soil. Therefore, sulphate fertilizer is usually most efficient when applied in the spring.

Elemental sulphur fertilizer applied in the spring of the crop year may not oxidize to sulphate sufficiently or early enough to supply the crop needs. Farmers who anticipate sulphur deficiencies and plan to utilize elemental S, should apply 10-20 lb/ac of elemental sulphur on an annual to biannual basis to avoid deficiencies.

POTASSIUM, CALCIUM AND MAGNESIUM

POTASSIUM

Potassium is a chemical element in the periodic table. It has the symbol K (L. *kalium*) and atomic number 19. The name "potassium" comes from the word "potash", as potassium was first isolated from potash. Potassium is a soft silvery-white metallic alkali metal that occurs naturally bound to other elements in seawater and many minerals. It oxidizes rapidly in air, is very reactive, especially in water, and resembles sodium chemically.

CALCIUM

Calcium is the chemical element in the periodic table that has the symbol Ca and atomic number 20. Calcium is a soft grey alkaline earth metal that is used as a reducing agent in the extraction of thorium, zirconium and uranium. Calcium is also the fifth most abundant element in the Earth's crust. It is essential for living organisms, particularly in cell physiology, and is the most common metal in many animals.

The most abundant isotope, ^{40}Ca, has a nucleus of 20 protons and 20 neutrons. Its electron configuration is: 2 electrons in the K shell (principal quantum number 1), 8 in the L shell (principal quantum number 2), 8 in the M shell (principal quantum number 3), and 2 in the N shell (principal quantum number 4). The outer shell is the valence shell, with 2 electrons in the lone 4s orbital, the 3p orbitals being empty.

Calcium is not naturally found in its elemental state. Calcium is found mostly in soil systems as limestone, gypsum and fluorite. Stalagmites and stalactites contain calcium carbonate. Being an essential macromineral in the human diet, soil conservation practices often consider the sustainable equilibrium of calcium concentrations in the earth.

All soils contain calcium ions (Ca_2+) and magnesium (Mg_2+) cations (positively charged ions) attracted to the negative exchange sites on clays and organic matter (cation exchange complex of the soil). The amount and relative proportion usually reflect the soil's parent materials. Calcium (Ca) and Mg are plant-essential nutrients, and the ionic form of each held on the soil exchange sites is the form taken up by plants. The usual approach for determining whether the soil supply is sufficient to meet crop needs is to extract soil with 1 molar (M) ammonium acctate (the same procedure used to determine soil test potassium) and evaluate the amount measured against critical levels. Because Iowa soils contain more than adequate levels of these nutrients, no critical level has been or can be established. Therefore, exchangeable Ca and Mg are not routinely tested, nor are there Iowa State University publications that give soil test Ca or Mg interpretations. Unless you are interested in the soil cation exchange capacity (CEC)—in routine soil testing determined by summing the dominant exchangeable cations (Ca_2+, Mg_2+, K+, H+)—soil samples don't need to be analyzed in the laboratory for Ca and Mg from most Iowa fields. Also, Iowa soils have large quantities of both nutrients and they are replenished by limestone application.

How is a Ca:Mg Ratio Calculated?

Once exchangeable Ca and Mg are determined by laboratory analysis, the ratio is calculated using the meq basis (electrical charge basis). For example, if there is 4.88 meq Ca/100 g soil and 1.72 meq Mg/100 g, then the Ca:Mg ratio is 2.8:1. Table gives the exchangeable Ca, Mg, and calculated Ca:Mg ratio for several Iowa soils. These values are typical for Iowa soils. Soil Ca:Mg ratios naturally are above 1:1.

Why the Interest in Ca:Mg Ratios?

From the above-mentioned statement that Ca and Mg levels are higher than needed for crop production in Iowa soils, you can easily conclude that ignoring the ratio is just fine. Research confirms that this conclusion is justified;

however, promotion of the ratio concept persists today despite many years of research that indicates otherwise.

The origin of this concept was derived from work by Bear and colleagues in the 1940s. However, their work did not differentiate between crop response (alfalfa) due to pH improvement from lime application to acid soils and the change in Ca:Mg. Other research during the same time indicated that ratios were not important. Many research trials since have not shown an influence of Ca:Mg ratio on crop production. Conclusions from the researchers were "The results strongly suggest that for maximum crop yields, emphasis should be placed on providing sufficient, but non-excesive levels of each basic cation rather than attempting to attain a favorable basic cation saturation ratio (BCSR), which evidently does not exist."

Various greenhouse and field trials indicate that crop productivity is not influenced by ranges from less than 1:1 to more than 25:1—ratios outside of what is normally measured in soils. Plants also play a role in Ca and Mg uptake and exclude excess Ca or Mg at the root surface.

In summary, the Ca:Mg ratio concept is unproven and should not be used as a basis for fertilization or liming practices. Having sufficient levels of Ca and Mg is the proper method of evaluation, rather than trying to manipulate ratios. We are fortunate in Iowa that soil Ca and Mg levels are normally adequate, and maintenance of plant-available Ca and Mg occurs either because the soil has a large supply or because of liming with local quarry limestone to maintain adequate soil pH for crop production.

MAGNESIUM

Magnesium is an essential ingredient of chlorophyll, the green plant pigment that gives leaves their colour and enables the plant to make food from sunlight.

Magnesium is essential for enzymes in plants and especially in chlorophyll – the green pigment that enables plant to carry out photosynthesis. Magnesium is a metallic element represented by the symbol Mg.

Soil pH

How well plants grow can be strongly influenced by whether the soil is acid or alkaline. It's best to get the pH right before dealing with other nutrient problems.

Lack of Magnesium

This appears as a lack of green in the leaves. The area between the veins of leaves, especially the lower older ones, goes yellow.

Sometimes the leaf margins also lose their green colour. Plants of the cabbage family sometimes develop a purplish colour. Sandy soils are most likely to lack magnesium. This deficiency is also seen when there is plenty of magnesium in the soil, but the roots cannot take it up because of poor soil structure, drought or water-logging. When the underlying problems are resolved the magnesium deficiency isappears.

Fast-growing plants may show a temporary magnesium deficiency in the older leaves. This usually sorts itself out when the plant begins to grow more slowly. There is no need to add extra magnesium. Magnesian limestone is acceptable to organic growers.

Common Magnesium Fertilizers

The amount of magnesium in fertilizer has to be stated on the packet. However, few compound fertilizers contain it, as plants need much less magnesium than nitrogen, phosphorus and potassium. Both Epsom salts (10 per cent Mg) and kieserite (16 per cent Mg) are fast-acting, water-soluble fertilizers containing magnesium sulphate.

Kieserite contains less water in its crystals than the morefrequently sold Epsom salts and is usually cheaper to use when treating large areas. Because it is less soluble, it can be used in seedbeds, whereEpsom salts could scorch delicate young roots.

Epsom salts are used for foliar feeding, as they are more soluble in water than ieserite. Neither fertilizer is organic.

Magnesium is easily washed out of soils that contain little or no clay. Since these soils are often acid and need lime, maintain the pH using magnesian limestone.

This ground limestone, derived from magnesium-rich rocks, is also known as Maglime or Dolodust. but do not use it where you intend to grow acid-loving plants like rhododendrons and heathers. Foliar-feed these plants instead – especially rhododendrons which can suffer from magnesium deficiency.

Some liquid fertilizers contain a little magnesium – these are a useful pick-me-up for containergrown plants – tomatoes in growing bags, for example. Manure, seaweed, compost and mushroom compost contain small amounts of magnesium. Cattle manure, for example, contains 0.04 per centMg.

Magnesium and Potassium

High levels of potassium (K) in the soil suppress plants' uptake of magnesium, so avoid overdosing with fertilizers containing potassium. Conversely, overfeeding with magnesium fertilizer can induce a potassium deficiency in plants. Foliar feeding helps overcome this imbalance.

General Guidelines

If your soil has not been analysedrecently and you believe that plants are showing magnesium deficiency, try a foliar feed in the first instance. If this has some effect, the soil may be deficient, and applying a magnesium fertilizer should be worthwhile. Spreading 30g a sq m (1oz a sq).

Suppliers

Magnesian limestone, Epsom salts and kieserite:

Bulky Organic Manures

Where the magnesium level is low or very low, manure, mushroom compost or garden compost are unlikely to provide enough and you will need fertilizers to boost levels for good growth. Where levels are high or very high, use good-quality manure or compost at 4.5kg a sq m (10lb a sq yd). Manures and composts are highly variable in composition, depending on what they have been made from, how they have been stored and how well-rotted they are. Also, they can vary greatly in water content. If in doubt, consider the organic matter as a soil conditioner and add fertilizer to provide nutrients.

Foliar Feeding

Wetting the leaves of plants suffering from a magnesium deficiency with a solution of Epsom salts can provide adequate magnesium. However, you may need to foliar feed again during the growing season and every time vulnerable plants are grown. Susceptible perennials, for example, typically need three or four applications per year.

Spraying young foliage during warm, humid weather gives the best results. Foliar feeding doesn't work so well on droughtstressed plants with sparsefoliage.

Avoid foliar feeding during hot, dry weather as the solution may dry too quickly, and the resulting concentrate could scorch the leaves.

If you're in doubt about the suitability of weather conditions, apply the feed in the evening.

Use 20g of Epsom salts per litre of water (3oz per gallon of water). Add a little washing-up liquid – this will act as a 'wetter' to stop the solution running straight off. Check if there is enough wetter by dipping a leaf in the solution – if it runs off without wetting the leaf, more is needed. Too much wetter can damage plants – if you're doing the whole of a precious plant, check by spraying a small shoot and leaving it for a few days before spraying the remainder.

Soil Analysis

To find out how much magnesium your soil contains requires soil analysis. This should give you a 'magnesium level' from very low to very high, or a 'magnesium index' from 0 to 3. You can then use the tables to determine what to add. Alternatively, a Gardening Which? Soil Analysis will tell you the levels and recommend what action to take.

TRACE ELEMENT

Trace element is a chemical element whose concentration is less than 1000 ppm or 0.1 per cent of rocks composition.

MICROMINERALS

Microminerals(also known as trace elements) are micronutrients that are chemical elements. They include boron, chromium, cobalt, copper, fluorine, iodine, iron, manganese, molybdenum, selenium, silicon, tin, vanadium, and zinc. They are dietary minerals needed by the human body in very small quantities (generally less than 100mg per day) as opposed to macrominerals which are required in larger quantities. Note that the use of the term "mineral" in this context is to be distinguished from the definition commonly used in the geological sciences.

MICRONUTRIENTS FOR PLANTS

There are about eight nutrients essential to plant growth and health that are only present in very small quantities. These are manganese, boron, copper, iron, chlorine, cobalt, molybdenum, and zinc. Some consider sulphur a micronutrient, but it is listed here as a macronutrient. Though these are present in only small quantities, they are all necessary.

- *Boron:* One of eight micronutrients. Boron is believed to be involved in carbohydrate transport in plants; it also assists in metabolic regulation. Boron deficiency will often result in bud dieback.
- *Chlorine*: One of eight micronutrients in plants. Chlorine is necessary for osmosis and ionic balance; it also plays a role in photosynthesis.
- *Cobalt*: One of eight micronutrients essential to plant health. Cobalt is thought to be an important catalyst in nitrogen fixation. It may need to be added to some soils before seeding legumes.
- *Copper*: One of eight micronutrients in plants. It is a component of some enzymes and of vitamin A. Symptoms of copper deficiency include browning of leaf tips and chlorosis.
- *Iron*: One of eight micronutrients in plants. It is essential for chlorophyll synthesis, which is why an iron deficiency results in chlorosis.
- *Manganese*: One of eight micronutrients. Manganese activates some important enzymes involved in chlorophyll formation. Manganese deficient plants will develop chlorosis between the veins of its leaves. The availability of manganese is partially dependent on soil pH.
- *Molybdenum*: One of eight micronutrients essential to plant health. Molybdenum is used by plants to reduce nitrates into usable forms. Some plants use it for nitrogen fixation, thus it may need to be added to some soils before seeding legumes.
- *Zinc*: One of eight micronutrients necessary for plant health. Zinc participates in chlorophyll formation, and also activates many enzymes. Symptoms of zinc deficiency include chlorosis and stunted growth.

Micronutrients are essential elements only needed by life in small quantities. Vitamins and trace minerals are sometimes included in the term.

Microminerals or trace elements include at least iron, cobalt, chromium, copper, iodine, manganese, selenium, zinc, and molybdenum. They are dietary minerals needed by the human body in very small quantities (generally less than 100mg/day) as opposed to macrominerals which are required in larger quantities.

Vitamins are organic chemicals that a given living organism requires in trace quantities for good health, but which the organism cannot synthesize, and therefore must obtain from its diet.

MACRONUTRIENTS

Macronutrients can be broken into two more groups: primary and secondary nutrients. The primary nutrients are nitrogen (N), phosphorus (P), and potassium (K). These major nutrients usually are lacking from the soil first because plants use large amounts for their growth and survival.

The secondary nutrients are calcium (Ca), magnesium (Mg), and sulphur (S). There are usually enough of these nutrients in the soil so fertilization is not always needed.

Also, large amounts of Calcium and Magnesium are added when lime is applied to acidic soils. Sulphur is usually found in sufficient amounts from the slow decomposition of soil organic matter, an important reason for not throwing out grass clippings and leaves.

Nitrogen (N)

Nitrogen is a part of all living cells and is a necessary part of all proteins, enzymes and metabolic processes involved in the synthesis and transfer of energy. Nitrogen is a part of chlorophyll, the green pigment of the plant that is responsible for photosynthesis.

Helps plants with rapid growth, increasing seed and fruit production and improving the quality of leaf and forage crops.

Nitrogen often comes from fertilizer application and from the air (legumes get their N from the atmosphere, water or rainfall contributes very little nitrogen).

Phosphorus (P)

Like nitrogen, phosphorus (P) is an essential part of the process of photosynthesis. Involved in the formation of all oils, sugars, starches, etc. Helps with the transformation of solar energy into chemical energy; proper plant maturation; withstanding stress.

Effects rapid growth. Encourages blooming and root growth. Phosphorus often comes from fertilizer, bone meal, and superphosphate.

Potassium (K)

Potassium is absorbed by plants in larger amounts than any other mineral element except nitrogen and, in some cases, calcium.

Helps in the building of protein, photosynthesis, fruit quality and reduction of diseases.

Potassium is supplied to plants by soil minerals, organic materials, and fertilizer.

Calcium (Ca)

Calcium, an essential part of plant cell wall structure, provides for normal transport and retention of other elements as well as strength in the plant. It is also thought to counteract the effect of alkali salts and organic acids within a plant.

Magnesium (Mg)

Magnesium is part of the chlorophyll in all green plants and essential for photosynthesis. It also helps activate many plant enzymes needed for growth.

Soil minerals, organic material, fertilizers, and dolomitic limestone are sources of magnesium for plants.

Sulphur (S)

- Essential plant food for production of protein.
- Promotes activity and development of enzymes and vitamins.
- Helps in chlorophyll formation.
- Improves root growth and seed production.
- Helps with vigorous plant growth and resistance to cold.

Sulphur may be supplied to the soil from rainwater. It is also added in some fertilizers as an impurity, especially the lower grade fertilizers. The use of gypsum also increases soil sulphur levels.

4

Fertilizers and Pesticides

INTRODUCTION

Agricultural chemicals fall into specific categories, depending on their purpose or method of application. An estimated 450 million pounds of pesticides were produced for use on the principle agricultural crops throughout the country in 1988. Fertilizers and pesticides are essential to modern agriculture, but if used incorrectly, they can cause economic loss, sickness and even death.

The skull and crossbones symbol is required by law to be on the label of all toxic pesticides, since the greatest hazards lie in the failure to follow precautions and directions for safe use, or in unexpected accidents such as the bursting of spray hoses, rusting containers, or improper storage and cleanup procedures.

When working with toxic chemicals, use respirators, gloves and eye protection as indicated by the manufacturer. Bathe and change clothes when you have completed the work.

A standard lawn or crop fertilizer is a relatively safe product. Multipurpose products such as lawn fertilizer and weed killer combinations are less toxic than other pesticides. Organophosphate pesticides destroy important enzymes in your nervous system without producing symptoms.

Ill effects also result from swallowing, breathing, or having skin or eye contact with pesticides. Skin absorption is the most common offender and is difficult to treat. Pesticides should be handled in well-ventilated areas and stored in their original, labeled containers—never in food or drink containers. Liquid pesticides should be stored in heated areas to prevent freezing.

Do not store volatile weedkillers or related herbicides with other "wettable" powders, dusts or seeds. Unidentified pesticides should be discarded, but never into a sewage system or street gutter that drains into open water, or into areas where soil, crops or animals will be contaminated. Never mix pesticides or drain a sprayer near a well, pond or stream, and be careful when burning empty pesticide containers—the smoke can be hazardous to all living organisms. To minimize the drift of pesticides, apply them on a calm day. Under windy conditions, the pesticide could blow or splash back into your face while you're mixing or pouring it.

FERTILIZER

Fertilizers or fertilizers are compounds given to plants with the intention of promoting growth; they are usually applied either via the soil, for uptake by plant roots, or by foliar spraying, for uptake through leaves. Fertilizers can be organic (composed of organic matter, *i.e.* carbon based), or inorganic (containing simple, inorganic chemicals). They can be naturally-occurring compounds such as peat or mineral deposits, or manufactured through natural processes (such as composting) or chemical processes (such as the Haber process).

Fertilizers typically provide, in varying proportions, the three major plant nutrients (nitrogen, phosphorus, and potassium), the secondary plant nutrients (calcium, sulphur, magnesium), and sometimes trace elements (or micronutrients) with a role in plant nutrition: boron, manganese, iron, zinc, copper and molybdenum.

INORGANIC FERTILIZERS (MINERAL FERTILIZER)

Examplcs of naturally-occurring inorganic fertilizers include Chilean sodium nitrate, mined "rock phosphate" and limestone (a calcium source, but mostly used to reduce soil acidity).

Examples of manufactured or chemically-synthesized inorganic fertilizers include ammonium nitrate, potassium sulfate, and superphosphate, or triple superphosphate.

MACRONUTRIENTS AND MICRONUTRIENTS

Fertilizers can be divided into macronutrients or micronutrients. There are three macronutrients: nitrogen, potassium, and phosphorus, which are consumed in high quantities and normally present as whole number percentages in plant tissues. There are many micronutrients, and their importance and occurrence differ from plant to plant. In general, most present from 5 to 100 parts per million (ppm) by mass. Examples of micronutrients are as follows: boron (B), calcium (Ca), copper (Cu), magnesium (Mg), iron (Fe), molybdenum (Mo), and zinc (Zn).

Macronutrient Fertilizers

Synthesized materials are also called artificial fertilizers, and may be described as straight, where the product predominantly contains the three primary ingredients of nitrogen (N), phosphorus (P) and potassium (K), which are known as N-P-K fertilizers or compound fertilizers when elements are mixed intentionally. They are named or labeled according to the content of these three elements, which are macronutrients. The mass fraction (per cent) nitrogen is reported directly. However, phosphorus is reported as diphosphorus pentoxide (P_2O_5), the anhydride of phosphoric acid, and potassium is reported as potash or potassium oxide (K_2O), which is is the anhydride of potassium hydroxide. Fertilizer composition is expressed in this fashion for historical reasons in the way it was analyzed (conversion to ash for P and K); this practice dates back to Justus von Liebig.

Consequently, an 18-51-20 fertilizer would have 18 per cent nitrogen as N, 51 per cent phosphorus as P_2O_5, and 20 per cent potassium as K_2O. Although analyses are no longer carried out by ashing first, the naming convention remains. If nitrogen is the main element, they are often described as nitrogen fertilizers.

Agricultural versus Horticultural Fertilizers. In general, agricultural fertilizers contain only one or two macronutrients. Agricultural fertilizers are intended to be applied infrequently and normally prior to seeding or concurrently with it. Examples of agricultural fertilizers are granular triple superphosphate, potassium chloride, urea, and anhydrous ammonia.

The commodity nature of fertilizer, combined with the high cost of shipping, leads to use of locally available materials or those from the closest/cheapest source, which may vary with factors affecting transportation by rail, ship, or truck. In other words, a particular nitrogen source may be very popular in one part of the country while a nother is very popular in another geographic region only due to factors unrelated to agronomic concerns.

Horticultural or specialty fertilizers, on the other hand, are formulated from many of the same compounds and some others to produce well-balanced fertilizers that also contain micronutrients. Some materials, such as ammonium nitrate, are used minimally in large scale production farming. The 18-51-20 example above is a horticultural fertilizer formulated with high phosphorus to promote bloom development in ornamental flowers.

Horticultural fertilizers may be water-soluble (instant release) or relatively insoluble (controlled release). Controlled release fertilizers are also referred to as sustained release or timed release. Many controlled release fertilizers are intended to be applied approximately every 3-6 months, depending on watering, growth rates, and other conditions, whereas water-soluble fertilizers must be applied at least every 1-2 weeks and can be applied as often as every watering if sufficiently dilute. Unlike agricultural fertilizers, horticultural fertilizers are marketed directly to consumers and become part of retail product distribution lines.

Justus von Liebig

Chemist Justus von Liebig (in the 19th century) contributed greatly to understanding the role of inorganic compounds in plant nutrition and devised the concept of Liebig's barrel to illustrate the significance of inadequate concentrations of essential nutrients. At the same time he deemphasized the role of humus. This theory was influential in the great expansion in use of artificial fertilizers in the 20th century.

Nitrogen fertilizer is often synthesized using the Haber-Bosch process, which produces ammonia. This ammonia is applied directly to the soil or used to produce other compounds, notably ammonium nitrate, a dry, concentrated product. It can also be used in the Odda Process to produce compound fertilizers such as 15-15-15.

Inorganic fertilizers sometimes do not replace trace mineral elements in the soil which become gradually depleted by crops grown there. This has been linked to studies which have shown a marked fall (up to 75 per cent) in the quantities of such minerals present in fruit and vegetables.[1] One exception to this is in Western Australia where deficiencies of zinc, copper, manganese, iron and molybdenum were identifed as limiting the growth of crops and pastures in the 1940's and 1950's. Soils in Western Australia are very old, highly weathered and deficient in many of the major nutrients and trace elements. Since this time these trace elements are routinely added to inorganic fertilizers used in Agriculture in this state.

In many countries there is the public perception that inorganic fertilizers "poison the soil" and result in "low quality" produce. However, there is very little (if any) scientific evidence to support these views. When used appropriately, inorganic fertilizers enhance plant growth, the accumulation of organic matter and the biological activity of the soil, while reducing the risk of water run-off, overgrazing and soil erosion. The nutritional value of plants for human and animal consumption is typically improved when inorganic fertilizers are used appropriately.

ORGANIC FERTILIZERS

Examples of naturally occurring organic fertilizers include manure and slurry, urine, peat, seaweed and guano. Green manure crops are also grown to add nutrients to the soil. Naturally occurring minerals such as mine rock phosphate, sulfate of potash and limestone are also considered Organic Fertilizers.

Examples of manufactured organic fertilizers include compost, dried blood, bone meal and seaweed extracts.

The decomposing crop residue from prior years is another source of fertility. Though not strictly considered "fertilizer", the distinction seems more a matter of words than reality.

Some ambiguity in the usage of the term 'organic' exists because some of synthetic fertilizers, such as urea and urea formaldehyde, are fully organic in the sense of organic chemistry. In fact, it would be difficult to chemically distinguish between urea of biological origin and that produced synthetically.

On the other hand, some fertilizer materials commonly approved for organic agriculture, such as powdered limestone, mined "rock phosphate" and Chilean saltpeter, are inorganic in the use of the term by chemistry.

Although the density of nutrients in organic material is comparatively modest, they have some advantages. For one thing organic growers typically produce some or all of their fertilizer on-site, thus lowering operating costs considerably. Then there is the matter of how effective they are at promoting plant growth, chemical soil test results aside. The answers are encouraging. Since the majority of nitrogen supplying organic fertilizers contain insoluble nitrogen and are slow release fertilizers their effectiveness can be greater than conventional nitrogen fertilzers.

Implicit in modern theories of organic agriculture is the idea that the pendulum has swung the other way to some extent in thinking about plant nutrition. While admitting the obvious success of Leibig's theory, they stress that there are serious limitations to the current methods of implementing it via chemical fertilization. They re-emphasize the role of humus and other organic components of soil, which are believed to play several important roles:

Mobilizing existing soil nutrients, so that good growth is achieved with lower nutrient densities while wasting less. Releasing nutrients at a slower, more consistent rate, helping to avoid a boom-and-bust pattern. Helping to retain soil moisture, reducing the stress due to temporary moisture stress

Improving the Soil Structure

Organics also have the advantage of avoiding certain long-term problems associated with the regular heavy use of artificial fertilizers:

The possibility of "burning" plants with the concentrated chemicals (*i.e.* an over supply of some nutrients) the progressive decrease of real or perceived "soil health", apparent in loss of structure, reduced ability to absorb precipitation, lightening of soil colour, etc. The necessity of reapplying artificial fertilizers regularly (and perhaps in increasing quantities) to maintain fertility the cost (substantial and rising in recent years) and resulting lack of independence. Organic fertilizers also have their disadvantages:

As acknowledged above, they are typically a dilute source of nutrients compared to inorganic fertilizers, and where significant amounts of nutrients are required for profitable yields, very large amounts of organic fertilizers must be applied. This results in prohibitive transportation and application costs, especially where the agriculture is practiced a long distance from the source of the organic fertilizer.

The composition of organic fertilizers tends to be highly variable, so that accurate application of nutrients to match plant production is difficult. Hence, large-scale agriculture tends to rely on inorganic fertilizers while organic fertilizers are cost-effective on small-scale horticultural or domestic gardens.

Finally, some organic fertilizers such as manures can contain bacteria or heavy metals harmful to human health, leading to a risk of the produce being contaminated and harmful to humans, or bearing disease.

In practice a compromise between the use of artificial and organic fertilizers is common, typically by using inorganic fertilizers supplemented with the application of organics that are readily available such as the return of crop residues or the application of manure.

It is important to differentiate between what we mean by organic fertilizers and fertilizers approved for use in organic farming and organic gardening by organizations and authorities who provide organic certification services. Some approved fertilizers may be inorganic, naturally occurring chemical compounds, *e.g.* minerals.

Environmental Effects of Fertilizer Use

Over-application of chemical fertilizers, or application of chemical fertilizers at a time when the ground is waterlogged or the crop is not able to use the chemicals, can lead to surface run-off (particularly phosphorus) or leaching into groundwater (particularly nitrates). Excessive surface run-off of easily soluble nutrients often pollutes lakes and streams is a process called eutrophication. This can lead to algal blooms which are destructive and even deadly to wildlife.

It is also possible to over-apply organic fertilizers. However: their nutrient content, their solubility, and their release rates are typically much lower than chemical fertilizers, partially because by their nature, most organic fertilizers also provide increased physical and biological storage mechanisms to soils.

The problem of over-fertilization is primarily associated with the use of artificial fertilizers, because of the massive quantities applied and the destructive nature of chemical fertilizers on soil nutrient holding structures. The high solubilities of chemical fertilizers also contribute to their tendency to rapidly pollute the world's ecosystems.

Storage and application of some fertilizers in some weather or soil conditions can cause emissions of the greenhouse gas nitrous oxide (N_2O). Ammonia gas (NH_3) may be emitted following application of inorganic fertilizers, or manure or slurry. Besides supplying nitrogen, ammonia can also increase soil acidity (lower pH, or "souring").

For these reasons, it is recommended that knowledge of the nutrient content of the soil and nutrient requirements of the crop are carefully balanced with application of nutrients in inorganic fertilizer especially. This process is called nutrient budgeting. By careful monitoring of soil conditions, farmers can avoid wasting expensive fertilizers, and also avoid the potential costs of cleaning up any pollution created as a byproduct of their farming.

The concentration of up to 100 mg/kg of Cadmium in phosphate minerals (for example Nauru[2] and the Christmas islands [3]) increases the contamination of soil with Cadmium, for example in New Zealand.[4] Uranium is another example for impurities of fertilizers

NITROGENOUS FERTILIZERS

According to the manner in which their nitrogen is combined with other elements, the nitrogenous fertilizers are divided into four groups; nitrate, ammonia, and ammonium salts, chemical compounds containing nitrogen in the amide form, and plant and animal by-products.

Inorganic Nitrogenous Fertilizers

Inorganic substances containing large amount of nitrogen come under this category. It is further divided into following groups according to the form of nitrogen they contain.

Nitrate fertilizers (NO_3^-): Nitrogen present in these fertilizers are in nitrate form, NO_3^- which are rapidly dissociated to release $NO3^-$ions and readily absorbed by the plants. Nitrate ions highly reactive and mobile are susceptible to losses due to leaching and under water-logged conditions by denitrification. They are alkaline in their residual effect in soil. Following are the nitrate fertilizers:

S. No.	Fertilizer	per cent N
1	Sodium nitrate ($NaNO_3$)	16%N
2	Calcium nitrate $Ca(NO_3)_2$	15.5%N

Ammonium fertilizers (NH_4^+): Ammonium fertilizers are soluble in water and, as such, absorbed on the soil colloids and thus protected from being washed away by run off or by leaching. Some crops like rice, sugarcane, tuber crop, seedlings directly utilise ammonium form of these fertilizers. The absorbed ammonium ions on soil collections are transformed to nitrate slowly and taken up by most of the crops. They are acidic in their residual effect in soil. Following are the ammonium fertilizers:

S. No.	Fertilizer	% N
1	Ammonium Sulphate –$(NH_4)_2SO_4$	20.6%N
2	Ammonium Chloride –NH_4CL	25%N
3	Ammonium phosphate—NH_4 (H_2PO_4)	20%N
4	Anhydrous ammonia —NH_3	82%N
5	Ammonia Solution —NH_3 in water	20 to 25%N

Nitrate and Ammonium fertilizers (Nitrate, $NO3^-$ and ammonium, NH_4^+): These fertilizers contain nitrogen in both nitrate ($NO3^-$) and ammonium forms (NH_4^+).

The nitrate nitrogen is readily available to plants for immediate need, whereas ammonium nitrogen becomes available to plants at a later stage, when it is transformed by microbiological process to nitrate. They are soluble in water and suitable for most of the crops and soils. They are acidic in its residual effect.

S. No.	Fertilizer	% N
1	Ammonium nitrate—NH_4NO_3	33 to 34%N
2	Calcium Ammonium Nitrate (CAN)$Ca(No_3)_2NH_4NO_3$	25%N3
3	Ammonium Sulphate Nitrate (ASN) $(NH_4)_2SO_4NH_4NO_3$	26%N

Organic Nitrogenous Fertilizers:

These fertilizers contain nitrogen in organic form. These include plant and animal by-products.

These fertilizers are relatively slow-acting but they supply nitrogen for a longer period.

Amide Fertilizers (Amine, NH2 or amide, CN2): These fertilizers contains nitrogen in organic compounds as amide—NH_2 or —— CN_2, not directly available to plants, as such, but quickly converted by soil microbes to ammoniacal and nitrate form and then utilised. Amide fertilizers are:

S. No.	Fertilizer	% N
1	Urea $CO(NH_2)_2$	46%N
2	Calcium cyanamide $CaCN_2$	21%N

Slow release nitrogenous fertilizers: These are newly developed fertilizers which release nitrogen in soil very slowly so that it may be available to the plants for longer period of time. Use of these materials result in better utilization of applied nitrogen by the growing crop plants and reduce losses.

S. No.	Fertiliser		% N
1	Urea-form (Urea+Formaldehyde)		38% N
2	Oxamide H_2NCO--$CONH_2$		31.8% N
3	Isobutylidine diurea (IBDU) (Urea+Isobutylaldehyde)$(CH_3)_2$---CH=CH – (NH---CO---$NH_2)_2$		32.2% N
4	Crotonilidine diurea (CDU)(Urea+acetaldehyde)		32% N
5	Guany1 urea (GU)		37% N
6	N-lignin (Ammonified lignin)		18% N
7	Sulphur coated urea (SCU)		36 to 40% N
8	Metal-ammonium phosphate Me.NH4PO4 x H2O		
	If Me is	Mg	8.3% N
	"	Fe	7.5% N
	"	Cu	7.2% N
	"	Zn	7.8% N
	"	Mn	7.5% N
	"	Co	6.1% N
9	Nutricate, Osmocate		Mixed fertilizers coated with various resin containing release controlling agents, additives

Phosphate Fertilizer Plants

Industry Description and Practices

Phosphate fertilizers are produced by adding acid to ground or pulverized phosphate rock. If sulphuric acid is used, single or normal, phosphate (SSP) is produced, with a phosphorus content of 16–21 per cent as phosphorous pentoxide (P2O5). If phosphoric acid is used to acidulate the phosphate rock, triple phosphate (TSP) is the result. TSP has a phosphorus content of 43–48 per cent as P2O5.SSP production involves mixing the sulphuric acid and the rock in a reactor. The reaction mixture

is discharged onto a slow-moving conveyorin a den. The mixture is cured for 4 to 6 weeks before bagging and shipping. Two processes are used to produce TSP fertilizers:run-of-pile and granular. The run-of-pile process is similar to the SSP process. Granular TSP uses lower-strength phosphoric acid (40 per cent,compared with 50 per cent for run-of-pile). The reaction mixture, a slurry, is sprayed onto recycled fertilizer

fines in a granulator. Granules grow and arethen discharged to a dryer, screened, and sent tostorage. Phosphate fertilizer complexes often have sulphuric and phosphoric acid production facilities.Sulphuric acid is produced by burning molten sulphur in air to produce sulphur dioxide, which is thencatalytically converted to sulphur trioxide for absorption in oleum. Sulphur dioxide can also be produced by roasting pyrite ore. Phosphoric acid is manufactured by adding sulphuric acid to phosphate rock. The reaction mixture is filtered to remove phosphogypsum, which is discharged to settling ponds or waste heaps.

Waste Characteristics

Fluorides and dust are emitted to the air from the fertilizer plant. All aspects of phosphate rock processing and finished product handling generate dust, from grinders and pulverizers, pneumatic conveyors, and screens. The mixer/reactors and dens produce fumes that contain silicon tetrafluoride and hydrogen fluoride. Liquid effluents are not normally expected from the fertilizer plant, since it is feasible to operate the plant with a balanced process water system. The fertilizer plant should generate minimal solid wastes. A sulphuric acid plant has two principal air emissions: sulphur dioxide and acid mist. If pyrites ore is roasted, there will also be particulates in air emissions that may contain heavy metals such as cadmium, mercury, and lead. Sulphuric acid plants do not normally discharge liquid effluents except where appropriate water management measures are absent. Solid wastes from a sulphuric acid plant will normally be limited to spent vanadium catalyst. Where pyrite ore is roasted, there will be pyrite residue, which will require disposal. The residue may contain a wide range of heavy metals such as zinc, copper, lead, cadmium, mercury, and arsenic.

The phosphoric acid plant generates dust and fumes, both of which contain hydrofluoric acid, silicon tetrafluoride, or both. Phosphogypsum generated in the process (at an approximate rate of about 5 tons per ton of phosphoric acid produced) is most often disposed of as a slurry to a storage/settling pond or waste heap. (Disposal to a marine environment is practiced at some existing phosphoric acid plants.)

Process water used to transport the waste is returned to the plant after the solids have settled out. It is preferable to use a closed-loop operating system, where possible, to avoid a liquid effluent.

In many climatic conditions, however, this is not possible, and an effluent is generated that contains phosphorus (as PO4), fluorides, and suspended solids. trace metals, fluorides, and radionuclides (especially radon gas) that have been carried through from the phosphate rock.

Pollution Prevention and Control

In a fertilizer plant, the main source of potential pollution is solids from spills, operating upsets, and dust emissions. It is essential that tight operating procedures be in place and that close attention be paid to constant cleanup of spills and to other housecleaning measurcs. Product will be retained, the need for disposal of waste product will be controlled, and potential contamination of stormwater run-off from the property will be minimized.

The discharge of sulphur dioxide from sulphuric acid plants should be minimized by using the double-contact, double-absorption process, with high efficiency mist eliminators.

Spills and accidental discharges should be prevented by using well-bunded storage tanks, by installing spill catchment and containment facilities, and by practicing good housekeeping and maintenance. Residues from the roasting of pyrites may be used by the cement and steel manufacturing industries.

In the phosphoric acid plant, emissions of fluorine compounds from the digester/reactor should be minimized by using well-designed, well-operated, and well-maintained scrubbers. Design for spill containment is essential for avoiding inadvertent liquid discharges. An operating water balance should be maintained to avoid an effluent discharge.

The management of phosphogypsum tailings is a major problem because of the large volumes and large area required and because of the potential for release of dust and radon gases and of fluorides and cadmium in seepage. The following measures will help to minimize the impacts:

- Maintain a water cover to reduce radon gas release and dust emissions.
- Where water cover cannot be maintained, keep the tailings wet or revegetate to reduce dust.
  (Note, however, that the revegetation process may increase the rate of radon emissions.)

- Line the tailings storage area to prevent contamination of groundwater by fluoride.
- Where contamination of groundwater is a concern, a management and monitoring plan should be implemented. Phosphogypsum may find a use in the production of gypsum board for the construction industry.

Target Pollution Loads

Implementation of cleaner production processes and pollution prevention measures can yield both economic and environmental benefits. The following production-related targets can be achieved by measures such as those described above. The numbers relate to the production processes before the addition of pollution control measures.

In sulphuric acid plants that use the double-contact, double-absorption process, emissions levels of 2–4 kilograms of sulphur dioxide per metric

ton (kg/t) of sulphuric acid can be achieved, and sulphur trioxide levels of the order of 0.15–0.2 kg/t of sulphuric acid are attainable.

Treatment Technologies

Scrubbers are used to remove fluorides and acid from air emissions. The effluent from the scrubbers is normally recycled to the process. If it is not possible to maintain an operating water balance in the phosphoric acid plant, treatment to precipitate fluorine, phosphorus, and heavy metals may be necessary. Lime can be used for treatment.

Spent vanadium catalyst is returned to the supplier for recovery, or, if that cannot be done, is locked in a solidification matrix and disposed of in a secure landfill. Opportunities to use gypsum wastes as a soil conditioner (for alkali soil and soils that are deficient in sulphur) should be explored to minimize the volume of the gypsum stack.

PESTICIDES

A pesticide as “any substance or mixture of substances intended for preventing, destroying, repelling, or mitigating any pest”.

A pesticide may be a chemical substance or biological agent (such as a virus or bacteria) used against pests including insects, plant pathogens, weeds, mollusks, birds, mammals, fish, nematodes (roundworms) and microbes that compete with humans for food, destroy property, spread disease or are a nuisance. Pesticides are usually, but not always, poisonous to humans.

TYPES OF PESTICIDES

- Bactericides for the control of bacteria,
- Herbicides for the control of weeds,

- Fungicides for the control of fungi,
- Insecticides for the control of insects,
- Miticides for the control of mites,
- Nematicides for the control of worms,
- Rodenticides for the control of rodents,
- Virucides for the control of viruses.

A systemic pesticide is a pesticide applied to a plant which is absorbed into its sap and so distributed throughout the plant to make all parts of it poisonous to pests, without harming the plant.

Bacteriocide

A bacteriocide or bactericide is a substance that kills bacteria and, preferably, nothing else.

Bacteriocidal antibiotics kill bacteria: bacteriostatic antibiotics only slow their growth or reproduction.

Herbicides

A herbicide is a pesticide used to kill unwanted plants. Selective herbicides kill certain targets while leaving the desired crop relatively unharmed. Some of these act by interfering with the growth of the weed and are often based on plant hormones. Herbicides used to clear waste ground are non-selective and kill all plant material with which they come into contact.

Classification of herbicides

Herbicides can be grouped by activity, use, chemical family, mode of action, or type of vegetation controlled.

By activity:

- Contact herbicides destroy only the plant tissue in contact with the chemical. Generally, these are the fastest acting herbicides. They are less effective on perennial plants, which are able to regrow from roots or tubers.
- Systemic herbicides are translocated through the plant, either from foliar application down to the roots, or from soil application up to the leaves. They can destroy a greater amount of plant tissue than contact herbicides.

By use:

- Soil-applied herbicides are applied to the soil and are taken up by the roots of the target plant.
- Pre-emergent herbicides are applied to the soil before the crop emerges and prevent germination or early growth of weed seeds.

- Post-emergent herbicides are applied after the crop has emerged. Their classification by mechanism of action (MOA) indicates the first enzyme, protein, or biochemical step affected in the plant following application. The main mechanisms of action are:
  - ACCase inhibitors are compounds that kill grasses. Acetyl coenzyme A carboxylase (ACCase) is part of the first step of lipid synthesis. Thus, ACCase inhibitors affect cell membrane production in the meristems of the grass plant. The ACCases of grasses are sensitive to these herbicides, whereas the ACCases of dicot plants are not.
  - ALS inhibitors: the acetolactate synthase (ALS) enzyme (also known as acetohydroxyacid synthase, or AHAS) is the first step in the synthesis of the branched-chain amino acids valine, leucine, and isoleucine. These herbicides slowly starve affected plants of these amino acids which eventually leads to inhibition of DNA synthesis. They affect grasses and dicots alike. The ALS inhibitor family includes sulfonylureas (SUs), imidazolinones (IMIs), triazolopyrimidines (TPs), pyrimidinyl oxybenzoates (POBs), and sulfonylamino carbonyl triazolinones (SCTs).
  - EPSPS inhibitors: The enolpyruvylshikimate 3-phosphate synthase enzyme EPSPS is used in the synthesis of the amino acids tryptophan, phenylalanine and tyrosine. They affect grasses and dicots alike. Glyphosate (Roundup) is a systemic EPSPS inhibitor but inactivated by soil contact.
  - Synthetic auxin inaugurated the era of organic herbicides. They were discovered in the 1940s after a long study of the plant growth regulator auxin. Synthetic auxins mimic this plant hormone. They have several points of action on the cell membrane, and are effective in the control of dicot plants. 2,4-D is a synthetic auxin herbicide.
  - Photosystem II inhibitors reduce electron flow from water to NADPH2+ at the photochemical step in photosynthesis. They bind to the Qb site on the D2 protein, and prevent quinone from binding to this site. Therefore, this group of compounds cause electrons to accumulate on chlorophyll molecules. As a consequence, oxidation reactions in excess of those normally tolerated by the cell occur, and the plant dies. The triazine herbicides (including atrazine) are PSII inhibitors.

Fungicides

A Fungicide is one of three main methods of pest control- chemical control of fungi in this case. Fungicides are chemical compounds used to prevent the spread of fungi in gardens and crops, which can cause serious damage to the plants. Fungicides are also used to fight fungal infections.

Fungicides can either be contact or systemic. A contact fungicide kills fungi when sprayed on its surface. A systemic fungicide has to be absorbed by the fungi before the fungi dies. One example is the QoI.

It is used in agriculture to avoid fungus growth because this can harm profits, Corn is often sprayed with fungicides and some fungicides are genitically modified to make sure that they get to work straight away by damaging all of the fungal cells.

In contexts other than agriculture, the term antifungal is often used for similar compounds.

Fungicides can also indirectly be harmful to human health as the cultivated crop is consumed and it can cause irritation as well as many symptoms such as headaches, diarrehea, damaged organs as well as severe disorders and maladies related to the nervous system. It is also a hazard to ecosystems as it can run off and can contaminate water bodies and as it bioaccumulates it can be increasingly toxic to living organisms in the ecosystem.

Insecticides

An insecticide is a pesticide used against insects in all development forms. They include ovocides and larvacides used against the eggs and larvae of insects. Insecticides are widely used in agriculture, household and even in medicine. Insecticides are very widely used in agriculture, as well as in people's dwellings and workplaces. The use of insecticides is one of the major factors behind the increase in agricultural productivity in the 20th century. Because insecticides have the potential to significantly alter ecosystems and may be concentrated through the food chain, it is necessary to balance agricultural needs with safety concerns when using them. The classification of insecticides is done in several different ways:

- The systematic insecticides which are incorporated into the plants and the insects take up the insecticide while feeding on the plants. The contact insecticides which take action when they are brought into direct contact with the insect by aerosol distribution.
- The natural insecticides which are made from plants as defence against insects like nicotine and pyrethrum. In contrast to the inorganic insecticides like arsenates copper- and fluorine compounds, which are now seldom used and the organic insecticides which is now the biggest group.
- The point of action is also a good way of sorting the insecticides. The neurotocic insecticides, the acetylcho-linesterase blocking and moulting blockers. The neurotoxic and coleterinas blocker are difficult because they are also toxic for other organisms. Where as the skinning

blockers and the other insecticides which take action only in metabolism of insects are less likely to be highly toxic on mammals.

- The contribution as aerosol (liquid or solid) is mostly used for contact insecticides, as solid for the uptake by the plant roots is preferred for systematic insecticides and fumigation with phosphine or hydrogen cyanide is mostly used for sensitive products after the harvest.

Miticides

Miticides are pesticides that kill mites. Antibiotic miticides, carbamate miticides, formamidine miticides, mite growth regulators, organochlorine, permethrin and organophosphate miticides are all in this category.

Acaricides: Methoprene is virtually harmless to non-insects.

Rodenticides

Rat poisons are a category of pest control chemicals intended to kill rats.

Single feed baits are chemicals sufficiently dangerous that the first dose is sufficient to kill.

Rats and certain other vermin are difficult to kill with poisons because their feeding habits reflect their place as scavengers. They will eat a small bit of something and wait, and if they don't get sick, they continue. A good rat poison must be tasteless and odorless in lethal concentrations, and have a delayed effect.

Metal phosphides

Metal phosphides have been used as a means of killing rodents. A mixture of food and a phosphide is left where the rodents can eat it.

Zinc phosphide is typically added to rodent baits in amount of around 0.75-2 per cent. The baits have strong, pungent garlic-like odour characteristic for phosphine liberated by hydrolysis. The odour attracts rodents, but has repulsive effect on other animals; birds, notably wild turkeys, are not sensitive to the smell, feed on the bait, and become collateral damage.

The tablets or pellets may also contain other chemicals which evolve ammonia which helps to reduce the potential for spontaneous ignition or explosion of the phosphine gas.

Phosphides used as rodenticides are:

- Aluminium phosphide,
- Aalcium phosphide,
- Zinc phosphide.

PLANT PROTECTION CHEMICALS

Protectants are "pesticidal substances" that are produced by plants to "protect" the plant from insects and other harmful organisms. Many plants incorporate protectants as a defence mechanism. Phytochemistry is in the strict sense of the word the study of phytochemicals. These are chemicals derived from plants. In a narrower sense the terms are often used to describe the large number of secondary metabolic compounds found in plants. Many of these are known to provide protection against insect attacks and plant diseases. They also exhibit a number of protective functions for human consumers.

PHYTOCHEMICALS

Phytochemicals are sometimes referred to as phytonutrients; these terms are often used interchangeably. In broad terms, they are said to be any chemical or nutrient derived from a plant source. However, in common usage, they have a more limited definition. They are usually used to refer to compounds found in plants that are not required for normal functioning of the body but that nonetheless have a beneficial effect on health or an active role in the amelioration of disease.

Thus, they differ from what are traditionally termed nutrients in that they are *not* a necessity for normal metabolism, and their absence will not result in a deficiency disease — at least not on the timescale normally attributed to such phenomena.

A minority claim that many of the diseases afflicting the people of industrialized nations are the result of those people's lack of phytonutrients in their diet. What is beyond dispute is that phytonutrients have many and various salubrious functions in the body. For example, they may promote the function of the immune system, act directly against bacteria and viruses, reduce inflammation, and be associated with the treatment and/or prevention of cancer, cardiovascular disease and any other malady affecting the health or well-being of an individual. Phytochemicals naturally occur in vegetables and fruit.

Families of Phytochemicals

The following are groups or families of related phytochemicals and common sources of phytochemicals arranged by family.

Family	Sources
Flavonoids	Berries, herbs, vegetables
Isoflavones (phytoestrogens)	Barley, flaxseed, soy
Isothiocyanates	Cruciferous vegetables
Monoterpenes	Citrus peels, essential oils
Organosulphur compounds	Chives, garlic, onions
Saponins	Beans, cereals, herbs
Capsaicinoids	Chile peppers
Phytosterols	Vegetable oils

Food Processing and Phytochemicals

Many phytochemicals are thought to be destroyed or removed by modern food processing techniques, possibly including cooking. For this reason, it is believed that industrially processed foods are less beneficial (contain fewer phytochemicals) than unprocessed foods. The absence or deficiency of phytochemicals is believed to have contributed to the increased prevalence of the above-cited preventable or treatable causes of death in contemporary society. Interestingly though, lycopene, which is a phytonutrient that can be found in tomatoes, is concentrated in processed foods such as spaghetti sauce and ketchup, making those foods much better sources of lycopene than fresh tomatoes.

Plant Protection

Since the beginning of agriculture pests have caused problems, but the coexistence of pest and crop plant had resulted in traditional farming systems that generally kept damage to a minimum. However, intensification, including changes of cultivars, crops and cultivation patterns, have often created conditions more favourable to pests. Migration or introduction of pests to new areas may also call for control where it was not previously needed. The classical example is grapes, where treatments are now required for diseases in Europe that were introduced from another continent in the last century.

Minimizing Pest Damage

Ways and means to protect crops from pests are many and varied. Numerous organisms have the capacity to become major pests but most of these are controllable by combinations of cultural measures, host plant resistance and biological control. Pesticides are used where these other control mechanisms do not keep pests below an acceptable level. For control of weeds, labour scarcity or cost has prompted a gradual shift from manual and mechanical means to use of chemicals.

Pesticide use has, however, led to a number of new problems: biological control agents may also be eliminated, or the pests may become resistant to the pesticide. Excessive or overly liberal use of pesticides may also harm human health and the environment. Lack of knowledge, risk-avoiding behaviour and aggressive sales techniques often result in a misuse or overuse of pesticides.

In promoting plant protection and pest management, FAO especially emphasizes the role of integrated pest management (IPM). One example of a recent successful project is the development and application of integrated pest control in rice growing in south and southeast Asia. Field trials were made to define appropriate IPM technology in farming areas of the participating countries. These trials involved the identification of natural enemies of the major insect pests, and the choice of effective methods of pest surveillance and of forecasting insect attack severity during the growing season. Such projects vary in the level of science and technology required or used. Some serve to establish national or regional research capabilities while others stimulate the extension of appropriate methodologies to farmers. A major FAO-executed programme presently under development is the application of IPM in vegetable production. The Organization's work on all aspects of IPM is directly related to sustainable agriculture.

In promoting plant protection and pest management, FAO especially emphasizes the role of integrated pest as the secretariat of the Convention and as an information exchange on pest distribution and pesticide legislation and regulations.

The 1989 Conference of FAO recommended that work be done on harmonization of the principles of plant quarantine legislation, of the principles of scientific evidence for plant quarantine measures, and of plant quarantine procedures.

PESTICIDES: A CODE FOR DISTRIBUTION AND USE

An important recent development has been the adoption by the FAO Conference (1985) of an International Code of Conduct on the Distribution and Use of Pesticides, amended in 1987 to include the principle of prior informed consent.

The pesticides industry, through the International Group of National Associations of Agrochemical Manufacturers (GIFAP), and UNEP in regard to the prior informed consent clause, agreed to assist FAO in the implementation of the provisions of the code. Environmental groups use the code as a yardstick to evaluate pesticide management in developing countries.

The code is supported by a set of guidelines prepared with the cooperation and advice of experts from FAO member countries. These guidelines cover such topics as registration and control (legislative and monitoring) of pesticides; efficacy data needed for registration; environmental criteria; residue data; labeling, packaging and storage; and disposal of waste pesticides and pesticide containers.

An activity complementary to the code is the provision of technical assistance to strengthen national and regional pesticide control schemes, to establish the necessary analytical facilities, and to give guidance on the safe and efficient use of pesticides.

SUSTAINING SMALL FARM ENTERPRISES

Agricultural development must seek to strike a balance between conservation of natural resources at national and local levels and meeting the needs of the country and the farming families, particularly over the short to medium term.

FARMING SYSTEMS

The key to sustainable development in low-resource areas is better management of farm production systems to minimize risks and to enable the farm household to withstand shocks and stress. In areas with adequate resources and a high potential for agricultural production, intensive crop production can generally be sustained using existing technologies so long as the soil's regenerative capacity is not exceeded.

Optimal production on land of high agricultural potential usually requires a steady supply of relatively expensive inputs, such as fuel, mineral fertilizers, herbicides and pesticides. Although sustainability and equity might be served by a shift on such land to mixed farming systems that require a lower level of external inputs, criteria of profitability even for small-scale producers and of ability to satisfy a growing demand might not be met. Thus any such change in farming system would have to be gradual. Available land with high potential for agricultural production is already cultivated, in some cases beyond its natural regenerative capacity and the financial capacity of the farmer to assist regeneration with fertilizer and other inputs. Increasingly, farmers are obliged to grow crops and livestock on land lacking one or more of the essential components for sustainable production.

Essential elements in resource-poor areas are economic and social incentives, community management of local projects, sound land-use planning that integrates use of forests and wooded areas, and improved farming systems that prevent or reverse land degradation.

A systems approach is especially needed in these more marginal areas to increase simultaneously the production of food, livestock feed and household fuel through closer integration of agriculture, forestry and fisheries, including aquaculture.

FAO places special importance on efficient operation of small farms in ways that are sustainable with available material and labour resources, that raise farm output, and that increase farm and farm-family income. On-site studies are made in cooperation with national institutions to characterize farm-household systems that are common in an area. Analysis of constraints on production efficiency or expansion is then used to identify feasible improvements which are tested and adapted in representative farming systems. Appropriate improvements are made available, along with training and guidance in farming system development, for extension by national or local staff of agricultural ministries or institutes.

Through the introduction and support of a wide range of engineering technologies FAO seeks to improve production, reduce the risks of failure, or alleviate drudgery for farm families, especially for women in developing countries. Governments, aid agencies and end-users are encouraged to select technology appropriate to the farming system in which it will be used.

AGROFORESTRY

Trees and shrubs serve several useful purposes, particularly in small-scale agriculture. In areas where the risk of erosion is high, they conserve and replenish the productive capacity of the soil. When integrated with production of crops and livestock in arid and semi-arid regions, trees can provide shade, fodder, fuelwood and numerous useful products. For example, in the Far East, farmers practising shifting cultivation have been shown how to plant exotic and improved local tree and bush species to provide forage, increasing livestock production as well as protecting the soil.

The importance of integrating tree growing in the various farming systems is often not recognized by policy-makers. Agroforestry should not be seen as a separate and isolated topic. Degraded lands can be rehabilitated, watershed management improved and sustainable farming systems made possible through agroforestry.

A promising agroforestry technique is alley cropping, in which crop plants occupy lanes 2 to 3 metres wide between lines of trees or shrubs-preferably leguminous, nitrogen-fixing species. When small ruminants are also part of the system, usually on a small scale, the technique is termed alley farming. In more arid areas, certain Acacia species may be the main source of fodder for animal production during the annual dry season and especially during extended droughts.

ENERGY

Energy is an indispensable input in agriculture. Much of the energy for plant production comes from the sun, as a gift for photosynthesis. But other inputs depend on different energy sources which are converted into fertilizers, weed killers, pesticides and aids to crop production and protection.

Agriculture is an important user of energy in rural areas. The energy inputs required by rural populations, however, are still in deficit in most developing countries. There are needs for heating, electricity or mechanical power for cooking food, transportation and other local services. More energy is needed but in many cases per caput energy consumption has declined with population growth,as a result of shortage of fuelwood supplies, transport problems and high costs.

Fuelwood accounts for about 60 per cent of the total energy consumption in Africa. About 30 developing countries depend on it for over two-thirds of their energy. Official statistics give an inadequate picture because national energy balances normally consider only the 'formal' demand for energy, and not the actual energy requirements of the rural population. The trend towards greater use of wood as fuel to meet urban needs must also be taken into account.

Modern technologies are available to generate energy in various forms, but they need to be adapted to local needs and possibilities. Biogas production, through anaerobic digestion of organic wastes and residues, is expanding in many developing countries. FAO supports the greater diffusion of the technology through national projects for both small and larger scale systems. Solar and wind energies are used for cooking, water heating, water pumping, solar greenhouses and refrigeration.

SUSTAINABLE AGRICULTURE AND RURAL DEVELOPMENT

Striking a balance between present needs for food, fuel and fibre, and the future requirements for a productive and healthy natural resources base is a major challenge facing agriculture today. The central principle of sustainable development is that the welfare of future generations should not be compromised by development and consumption patterns of the present generation.

FAO has defined sustainable development as "the management and conservation of the natural resource base, and the orientation of technological and institutional change in such a manner as to ensure the attainment and continued satisfaction of human needs for present and future generations. Such sustainable development (in the agriculture, forestry and fisheries sectors) conserves land, water, plant and animal genetic resources, is environmentally non-degrading, technically appropriate, economically viable and socially acceptable."

5

Water Resources

INTRODUCTION

Integrated water resources management can take different forms and is examined best in specific situations. In the water-supply field, the term "integrated resource planning" has come into use to express concepts of integration in supply development. Perhaps the most comprehensive concept for water supply is "Total Water Management."

According to a 1996 report of the American Water Works Research Foundation, Total Water Management is the exercise of stewardship of water resources for the greatest good of society and the environment. A basic principle of Total Water Management is that the supply is renewable, but limited and should be managed on a sustainable-use basis. Taking into consideration local and regional variations, Total Water Management:

- Encourages planning and management on a natural water systems basis through a dynamic process that adapts to changing conditions;
- Balances competing uses of water through efficient allocation that addresses social values, cost effectiveness and environmental benefits and costs;
- Requires the participation of all units of government and stakeholders in decision-making through a process of coordination and conflict resolution;
- Promotes water conservation, reuse, source protection and supply development to enhance water quality and quantity;
- Fosters public health, safety and community goodwill.

This definition focuses on the broad aspects of water supply. Examples can be given for other situations, including water-quality management planning, water allocation and flood control.

WATER RESOURCES

HYDROLOGIC IMPLICATIONS OF CLIMATE

Climate change is likely to alter the hydrologic cycle in ways that may cause substantial impacts on water resource availability and changes in water quality. For example, the amount, intensity and temporal distribution of precipitation are likely to change. Less dramatic but equally important changes in run-off could arise from the fact that the amount of water evaporated from the landscape and transpired by plants will change with changes in soil moisture availability and plant responses to elevated CO_2 concentrations. This will affect stream flows and groundwater elevations. This overview briefly summarizes potential impacts on the most important water resource elements.

PRECIPITATION CHANGES

Along with the projected future warming there will be changes in atmospheric and oceanic circulation, and in the hydrologic cycle, leading to altered patterns of precipitation and run-off. The most likely will be an increase in global average precipitation and evaporation as a direct consequence of warmer temperatures. Evaporation will increase with warming because a warmer atmosphere can hold more moisture and higher temperatures increase the evaporation rate. On average, current climate models suggest an increase of about 1 per cent–2 per cent per degree Celsius from warming forced by CO_2. An increase in global average precipitation does not mean that it will get wetter everywhere and in all seasons. In fact, all climate model simulations show complex patterns of precipitation change, with some regions receiving less and others receiving more precipitation than they do now; changes in circulation patterns will be critically important in determining changes in local and regional precipitation patterns.

CHANGES IN PRECIPITATION FREQUENCY AND INTENSITY

Many have argued that, in addition to changes in global average precipitation, there could be more pronounced changes in the characteristics of regional and local precipitation due to global warming. For example, Trenberth *et al.* hypothesized that, on average, precipitation will tend to be less frequent, but more intense when it does occur, implying greater incidence of extreme floods and droughts, with resulting consequences for water storage. Thus, the prospect may be for fewer but more intense rainfall–or snowfall–events.

CHANGES IN AVERAGE ANNUAL RUN-OFF

Run-off changes will depend on changes in temperature and precipitation, among other variables. Arnell used several climate models to simulate future climate under differing emissions scenarios. The study linked these climate simulations to a large-scale hydrological model to examine changes in annual average surface run-off by 2050.

They found that all simulations yield a global average increase in precipitation, but likewise exhibit substantial areas where there are large decreases in run-off. Thus, the global message of increased precipitation clearly does not readily translate into regional increases in surface and groundwater availability.

HYDROLOGICAL IMPACTS ON COASTAL ZONES

The IPCC Working Group II TAR identifies several key impacts of sea level rise on water providers located in coastal areas, including:

- Lowland inundation and wetland displacement,
- Altered tidal range in rivers and bays,
- Changes in sedimentation patterns,
- More severe storm surge flooding,
- Increased saltwater intrusion into estuaries and freshwater aquifers, and
- Increased wind and rainfall damage in regions prone to tropical cyclones.

WATER QUALITY CHANGES

Where stream flows and lake levels decline, water quality deterioration is likely as nutrients and contaminants become more concentrated in reduced volumes. Warmer water temperatures may have further direct impacts on water quality, such as reducing dissolved oxygen concentrations. Prolonged droughts also tend to allow accumulation of contaminants on land surfaces, which then pose greater risks when precipitation returns.

At the other extreme, heavy precipitation events may result in increased leaching and sediment transport, causing greater sediment and non-point source pollutant loadings to watercourses. Floods, in particular, increase the risk of water source contamination from sewage overflows, agricultural land, and urban run-off.

WATER STORAGE AND MANAGEMENT

An intensified hydrological cycle could make reservoir management more challenging, because there is often a trade-off between storing water for dry period use and evacuating reservoirs before the onset of the flood season to protect downstream communities. Reservoirs have been usually sized to handle a certain amount of stream flow variability, determined from a relatively short historical record. If the variability increases, reservoirs may be undersized to meet demands or adequately serve as flood protection.

Thus, it may become more difficult to meet delivery requirements during prolonged periods between reservoir refilling without also increasing the risk of flooding. Earlier spring run-off from snowmelt is a likely manifestation of global warming. To the extent that adequate reservoir space is available, changing the operation procedures of reservoirs could mitigate some of these effects.

GROUNDWATER CHANGES

In many communities, groundwater is the main source of water for both irrigation and municipal and industrial demands. Generally there are two types of groundwater resources–renewable and non-renewable. Renewable groundwater is directly tied to near-surface hydrologic processes; it is thus intricately tied to the overall hydrologic cycle and could be directly affected by climatic change.

In many places, the overdraft of renewable groundwater aquifers occurs because the rate of withdrawal exceeds the rate of recharge. In fact, renewable groundwater supplies are often thought of being the same resource as surface water because they are so intertwined.

Thus, climate changes could directly affect these recharge rates and the sustainability of renewable groundwater. Non-renewable groundwater supplies are usually derived from deep earth sediments deposited long ago and so have little climatic linkage.

WATER DEMAND CHANGES

Future climate change could affect municipal and industrial water demand. Municipal water demand–especially for garden, lawn, and recreational field watering–is affected by climate to a certain extent, but rates of water use are highly dependent on water resource regulations and local user education. Industrial use for processing purposes is relatively insensitive to climate change; it is conditioned by technologies and modes of use. Demand for cooling water would be affected by a warmer climate because increased water temperatures will reduce the efficiency of cooling, perhaps necessitating increased source water abstraction to meet cooling requirements (or, alternatively, changes in cooling technologies to make them more efficient).

REGIONAL CHANGES

Although the preceding sections postulate some expected hydrologic changes from global warming, these generalizations will not be applicable in all places and at all times. Watson *et al.* examined the regional impacts of climate change, with a particular focus on assessing vulnerability. The report noted that more than 1 billion people do not have access to adequate water supplies, and that some 19 countries, primarily in the Middle East and Northern and Southern Africa, face severe water shortfalls. This number could double by 2025, in large part because of the increased demand caused by economic and population growth. Climate change could exacerbate the situation.

Watson *et al.* noted that many developing countries are particularly vulnerable to climate change because they already experience water shortfalls, being in arid and semi-arid regions. Many people derive their water from single-point systems, such as boreholes or isolated reservoirs.

This lack of "water diversification" increases people's vulnerability to water shortage. These systems do not have the redundancy necessary to minimize the risks during times of shortage. Also, given the often limited technical, financial and management resources available to developing countries, adjusting to shortages or implementing adaptation measures can impose a burden on national economies.

These small water supply systems are found in many parts of the world. Persistent drops in water level in these systems could adversely affect the quality of water by increasing the concentrations of sewage waste and industrial effluents, thereby increasing the potential for outbreaks of disease and reducing the quantity of potable fresh water available for domestic use.

There is evidence that flooding is likely to become a larger problem in many temperate and humid regions, requiring adaptation not only to droughts and chronic water shortages but also to floods and associated damage, raising concerns about dam safety.

Trenberth *et al.* hypothesized that global warming due to enhanced GHGs could increase the intensity of precipitation and reduce its frequency, which would be particularly problematic in regions with rapid changes in land use and land cover, because this would mean changes in surface run-off and groundwater recharge characteristics.

Flooding could be worse, accompanied by rapid drying and less overall water resource availability. The effects on water resources could be sufficient to lead to conflicts among users, regions and countries.

NATURAL RESOURCES OF WATER

Water is a natural resource critical to the environment. Water also is an economic resource critical to society. Unfortunately, people who champion water as an economic resource essential to society commonly see themselves in direct opposition to those who champion water as a critical component of the environment that must be conserved and protected. This creates a difficult situation for those who must manage the available water resources to best satisfy all of the diverse interests that depend on water.

It is, of course, impossible to successfully manage water resources solely from either of the two extremes. If public policy exclusively protected ecological interests at the expense of economic interests, then it could threaten society's ability to meet the basic need all people have for water.

From an ecological perspective, this could weaken government and diminish the ability to protect the environment, whereas from an economic standpoint, jobs and essential services such as housing and communications could be lost. Likewise, if water management policies focus entirely on the economy, then the environment often is harmed. A diminished environment affects the public health, the quality of life and ultimately the ability to survive on planet Earth. Clearly, society is as dependent on a healthy environment as it is on a healthy economy.

LEGISLATION

In many Western Hemisphere countries, water resources are managed through legislation and regulations. In the United States, elected officials in Congress and the state legislatures set the policies that prioritize where water can or cannot be used and how much water must remain to function in the natural environment. These laws are then implemented through regulatory agencies at the federal, state and local levels of government. One example of this legislation is the federal Clean Water Act, which requires that the quality of surface-water bodies, such as rivers and lakes, be kept at a level that will assure the maintenance of both environmental and economic functions.

Public Opinion

The effectiveness of these policies to meet both environmental and economic interests is dependent, in part, on how well the laws are written and how effectively they are enforced. However, public opinion strongly affects the type of policies that get enacted. For example, in the 1930s, the United States was emerging from a severe economic depression, so the growth and protection of the economy were paramount in the minds of the public and policymakers alike. As a result, the national policy from roughly 1930 to 1970 was to use the water of major rivers to provide jobs and to allow cities to grow in order to expand the economy. This was often accomplished through the federal government's dam-building projects for hydroelectric power and water supply, particularly for irrigation.

Shifting Sentiment

The dam-building policy was beneficial to the economy, but by 1970 national policy and public sentiment had shifted towards the environmental health of rivers and lakes, many of which had been harmed by the dam projects. One outgrowth of this environmental pendulum shift was the Endangered Species Act (ESA) passed by the U.S. Congress in 1973.

The ESA is an example of legislation that fundamentally changed water policy without directly addressing water issues. On its face, the law would not seem to be about water policy. When applied to fish species, however, it places requirements on water resource management to meet certain stream flow levels in order to protect a given species.

Though this is clearly a good policy, it often requires extraordinary efforts in the regional management of water to ensure that the water demands of the society are met without further endangering the listed species. These efforts can be very costly. A balance must be found whereby the environment is protected to maintain the function of the rivers and lakes while still meeting the water demands of the community.

CLIMATE CHANGE INFORMATION FOR WATER RESOURCE MANAGEMENT

Water managers are accustomed to adapting to changing circumstances, many of which may be analogues of future climate change, and they have developed a wide range of adaptive options that may or may not be appropriate, depending on the magnitude of climate change and how rapidly and when those changes are put into affect. For reasons noted above, climate warming will inevitably challenge existing water management practices, especially in countries with less experience in incorporating uncertainty into water planning and less financial and institutional resources. The current challenge is to incorporate climate change uncertainty along with the other types of uncertainty usually treated in the water planning process.

A cornerstone of climate change analysis in the water planning process is the use of hydrologic simulation to study the effect of a changing climate on the rainfall–run-off process. Many of these models attempt to capture the physical mechanisms of run-off production across the landscape by characterizing precipitation on to the land surface, either directly or through snowmelt, and the partitioning of that water into evapotranspiration, run–off to the river network, and recharge to groundwater systems.

Integrated Water Resource Management (IWRM) models can then use these water fluxes to determine reservoir management and water delivery strategies, often within a well-defined regulatory framework.

A review of the scientific and water planning literature, however, suggests that most water resource studies on climate change have incorporated climate information into their planning process using a top-down approach. This approach typically begins by establishing the scientific credibility of human-caused climate warming, develops future climate scenarios to be used at the regional level, and then imposes those potential changes on water resource systems to assess, for example, system reliability. The problems with a top-down approach are as follows:

- It does not always address the unique needs of a region, and
- The approach may become mired in the uncertainty of the future climate projections.

Alternatively, the bottom-up approach begins by identifying a water sector's most critical vulnerabilities; articulates the causes for those vulnerabilities; suggests how climate change, climate variability, and climate extremes might or might not exacerbate those vulnerabilities; and finally designs an analytic process to better address and solve the vulnerability in the face of the climatic uncertainty (*e.g.*, a precautionary approach).

In either top-down or bottom-up approaches, IWRM can be the most effective method for assessing adaptation options and their implications in the context of an evolving regulatory environment with its competing demands.

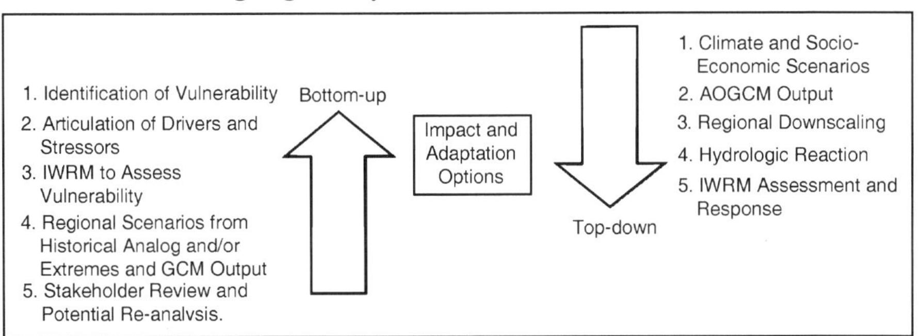

Fig. Top-Down vs. Bottom–Up Approaches to Climate Change Assessment.

ADAPTATION TO CLIMATE CHANGE IN THE WATER RESOURCE SECTOR

Water managers have long had to cope with the challenges posed by hydrologic variability, essentially adapting to variability. These adaptations have included the development of coupled reservoir and irrigation systems that allow for the redistribution of water during wet seasons and-normal water years, for carry-over use during dry seasons or to minimize the impacts of drought. Other adaptations, to minimize extensive flooding due to heavy precipitation and high flows, have been the creation of dams and levees to protect cities and concentrated agricultural production.

Protection of growing communities from the risk of flooding has increased in importance in many places. Two adaptation strategies have been the development of flood control operating rules for large reservoirs and the construction of flood bypasses and levees.

These adaptations have had profound effects on ecosystems in many places, and the response has been the establishment of minimum instream flow requirements at important points in watersheds. There is also a growing recognition that ensuring the proper volume of flow for ecosystems is necessary but not sufficient.

Other factors, such as the temperature and quality of the water in rivers, are also important. Recent adaptations with regard to water temperature have included the construction of temperature control devices in large dams that allow for the controlled management of cold and warm water pools which generally develop when large reservoirs stratify.

Water quality adaptation measures can include the development of discharge permitting requirements. These have been limited, to date, to point discharges, but are now being contemplated for non-point sources as well.

WATER RESOURCE MANAGEMENT METHOD

The IWRM method is a systematic approach to planning and management that considers a range of supply-side and demand-side processes and actions, and incorporates stakeholder participation in decision processes. It also facilitates adaptive management by continually monitoring and reviewing water resource situations.

To capture the supply and demand side processes and actions, IWRM must simultaneously address the two distinct systems that shape the water management landscape.

Factors relating to the biophysical system shape the demand for water (through pricing, incentives for water reclamation and recycling, demand management programmes, etc.), availability of water, and its movement through a watershed; factors relating to the socio-economic management system shape how available water is stored, allocated, regulated and delivered within or across watershed boundaries.

Increasingly, operational objectives of the management system seek to balance water for human use and water for environmental needs. Thus, integrated analysis of the natural and managed systems is arguably the most useful approach.

Groundwater Wells/Groundwater Fluxes

This type of analysis relies on hydrologic modelling tools that simulate physical processes, including precipitation, evapotranspiration, run-off, infiltration and groundwater flow.

In managed systems, analysts must also account for the operation of hydraulic structures, such as dams and diversions, as well as institutional factors that govern the allocation of water between competing demands, including consumptive demand for agricultural or urban water supply or non-consumptive demands for hydropower generation or ecosystem protection.

Because water quality also will change with climate, special attention must also be paid to water quality changes. Such changes may result in increased restrictions on water withdrawals to maintain water quality and ecosystem health. Changes in each of these elements can influence the ultimate impacts of climate change on water resources.

Although different hydrologic models can yield different values for stream flow, groundwater recharge, water quality results, etc., their differences have historically been small compared to the uncertainties attributed to global warming reflected in the differences among GCM output.

The chain of effects, however, from climate to hydrologic response, to water resource systems to the actual impacts on water supply, power generation, navigation, water quality, etc., will depend on many factors, each with a different level of uncertainty.

Currently, infrastructure investments and long-term management strategies assume that precipitation and run-off will follow past trends. Mounting evidence for climate change makes this an increasingly tenuous assumption.

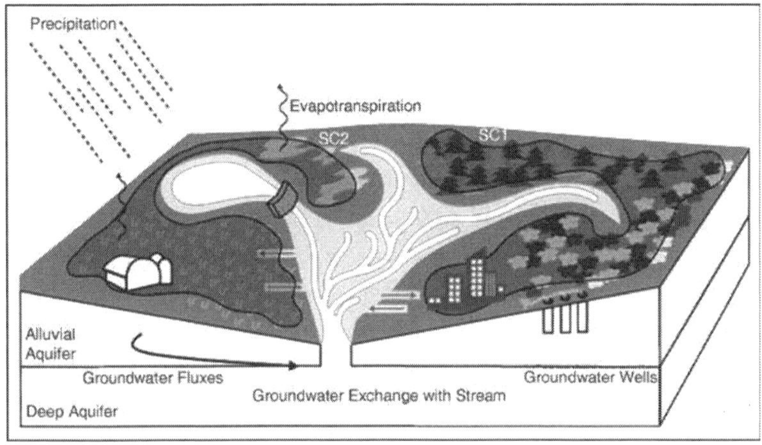

Fig. The Interface between the Natural Watershed and Managed System.

INTEGRATED WATER RESOURCE MANAGEMENT MODELS

The water management literature is rich with IWRM models that have focused either on understanding how water flows through a watershed in response to hydrologic events or allocating the water that becomes available in response to those events. Hydrologic simulation attempts to capture the most important land–atmosphere components of the hydrologic cycle. One well-known hydrologic simulation tool is the United States Department of Agriculture's Soil Water Assessment Tool.

The SWAT model has sophisticated physical hydrologic watershed modules that describe, among other things, rainfall–run-off processes, irrigated agriculture processes, and point and non-point watershed dynamics.

The Danish Hydraulic Institute (DHI) offers a suite of models, including MIKE SHE, which is an integrated water resource modelling tool that can simulate all the major processes in the land phase of the hydrologic cycle. The United States Army Corp of Engineers Hydrologic Engineering Centre (HEC) developed the HEC-Hydrologic Modeling System (HMS), which simulates the precipitation–run–off processes of dendritic watershed systems. The Delft Hydraulics Laboratory developed the HYMOS model for surface and groundwater hydrology and includes rainfall–run–off simulation.

Important to mention, but perhaps of less importance in terms of water supply and demand management, are hydrodynamics models. These models are typically developed to track the propagation of water through a river system at very short time steps (*e.g.*, minutes to hours). Applications include flood inundation mapping and flood forecasting. From a climate change perspective, this sort of hydrodynamics model could be used to study the interacting effects of sea level rise and changes in freshwater discharge into bay–delta systems at the event scale.

Models of this type include Delft's DELFT3D, a sophisticated two- and three-dimensional hydrodynamics model that can simulate non-steady flow and transport resulting from tidal and meteorological forcing. Other hydrodynamics models are DHI's MIKE21 and MIKE3, which can simulate hydraulics, water quality and sediment transport in rivers, lakes and coastal areas, and can be used for local to regional watershed studies. The HEC's hydrodynamics models include HEC–RAS, which is a one-dimensional model for hydraulic calculations and water surface profiles.

This CGE training material focuses more attention on water resource management modelling, which can be used to investigate water supply and demand issues over long planning time horizons, consistent with climate change projections. Water resource management models include the RiverWare DSS, a state–of–the–art water planning and operations model that can be used to develop multiobjective simulations and optimizations of river and reservoir systems, such as and including storage and hydropower reservoirs, river reaches, diversions and water users.

The DHI modelling group provides a GIS-based planning model for studying water management options, referred to as MIKE BASIN. The HEC-ResSim is a reservoir simulation model that can describe operating rules, such as release requirements and constraints, hydropower requirements and multiple reservoir operations, but it too requires prescribed flows from other models. The MODSIM DSS is a generalized river basin network flow model that can simultaneously incorporate the complex physical, hydrological and institutional/administrative aspects of river basin management, including water rights, but boundary flows must be prescribed.

The MULINO DSS is a multisectoral, integrated and operational decision support system for sustainable use of water resources at the catchment scale, with a focus on the DSS as a multicriteria decision aid. Similar to RiverWare, MULINO can accommodate a physical watershed hydrology model that is external to the system, linked through appropriate input–output procedures. WaterWare is a sophisticated water resource DSS that includes dynamic simulation of physical models of water quality, allocation, rainfall–run–off, groundwater and water management elements, including demand/supply, cost–benefit analysis and multicriteria analysis. Although WaterWare integrates the physical hydrology and the management system, applying the model requires a rather sophisticated level of user and hardware support. The Delft Hydraulics River Basin Simulation Model (RIBASIM) is a generic water resource planning model for investigating the behaviour of river basins under various hydrologic conditions. The WEAP21 attempts to address the gap between water management and watershed hydrology, and the requirements that an effective IWRM be useful, easy to use, affordable and readily available to the broad water resource community. WEAP21 integrates a range of physical hydrologic processes with the management of demands and installed infrastructure in a seamless and coherent manner. It allows for multiple scenario analysis, including alternative climate scenarios and changing anthropogenic stressors, such as land use variations, changes in municipal and industrial demands, alternative operating rules and points of diversion changes. WEAP21's strength is in addressing water planning and resource allocation problems and issues, and, importantly, it is not designed to be a detailed water operations model that might be used to optimize hydropower based on hydrologic forecasts.

The management system in the WEAP21 DSS is described by a user-defined demand priority and supply preference set for each demand site used to construct an optimization routine that allocates available supplies. Demands are defined by the user, but typically include municipal and industrial demand, irrigated portions of subcatchments, and instream flow requirements. Demand analysis in WEAP that is not covered by the evapotranspiration-based irrigation demand follows a disaggregated, end-use-based approach for determining water requirements at each demand node. Economic, demographic and water use information is used to construct alternative scenarios that examine how total and disaggregated consumption of water evolve over time. These demand scenarios are computed in WEAP and applied deterministically to a linear-programme-based allocation algorithm. Demand analysis is the starting point for conducting integrated water planning analysis because all supply and resource calculations in WEAP are driven by the optimization routine that determines the final delivery to each demand node, based on the priorities specified by the user.

USING IWRM FOR ADAPTATION ASSESSMENT

Water managers have long had to cope with the challenges posed by climate and hydrologic variability, both intra–annually and inter-annually. Their adaptation strategies include responding to both seasonal variability and extended wet and drought periods using integrated reservoir and irrigation systems that allow for the capture of water during the wet season for use during dry seasons and extended drought periods. Other adaptations have been the use of levees and dams in concert to protect communities from heavy precipitation and high flood flows during extreme wet periods. Climate change might challenge these conventional adaptations, forcing more aggressive strategies, such as artificially recharging aquifers or developing integrative strategies that optimally operate reservoirs in conjunction with the artificial recharging of aquifers.

Protecting growing communities from the risk of flooding also has increased in importance. Two adaptation strategies are flood control operating rules for large reservoirs and flood bypasses that move water away from human settlements. In many cases, flood conveyance restricts land use to not allow permanent structures, so during periods of high flow, large volumes of water can be diverted from the main river channels, thereby reducing the risk of flooding along the developed riverfront areas. Climate change adaptation could include the re–operation of reservoirs to maintain their important services and, in some cases, the construction of new facilities to help in flood protection or to secure water supplies. These adaptations would require careful consideration to ensure their usefulness in adapting to climate change, with an IWRM modelling process a key in assessing the benefits of alternative strategies for adaptation to climate change. Other water resource adaptations are designed to limit impacts on important aquatic ecosystems. Adaptations include establishing minimum instream flow requirements at important points in the water system. The physical rehabilitation of riverine ecosystems has taken on a priority because the watershed itself can act as a flow regulating system, buffering rapid stream flow response as water slowly migrates through the complex pathways of the watershed.

Water planners now realise that the extensive channelization of many watersheds around the world limits the amount of wetland and riparian habitat available. Setting communities back from the riverine corridor is an adaptation being considered together with the concept that flow-bypass structures can be managed as wetland complexes. There is also growing recognition that ecosystems require more than the minimum flows often prescribed for them.

Other factors, such as the temperature and quality of the water in rivers, are also important. Adaptations with regard to water temperature include the construction of temperature control devices in large dams which allow for the controlled management of cold and warm water pools that generally develop when large reservoirs stratify, which could be amplified with climate change. Water quality adaptation measures can include the development of discharge permitting requirements. To date these have been limited to point discharges, but they are now being contemplated for non–point sources as well. There has been a clear historical trend towards placing a higher priority on environmental security as societies place greater value on the role of ecosystems.

DATA

Water resource planning models require data on water demand and water supply. Water demand information usually needs to come from local sources– including water use per capita domestic rates, and industrial and commercial water use rates. Common water use rates can be extracted from literature. Irrigation demands can be determined from knowledge about crops and the climate. Data on demand for cooling water for thermal power plants and mainstream demands for navigation, recreation and hydropower are usually available from the users. Data on ecosystem demand may be available from environmental agencies. All the demands must be adjusted for climate change.

WATER MANAGEMENT AND DEMANDS OF ENVIRONMENTAL LAWS

Even without the demands of environmental laws, minimizing impacts on streams and rivers while maintaining water supplies for cities and industry is a complex task because of natural constraints on water availability. Areas of the United States like the arid Southwest have limited amounts of available water such that any additional use has the potential to significantly harm natural systems. However, water resource management is difficult even in areas with substantial annual rainfall. Here the problem is often one of timing because water may be abundant only in the winter months when water demand is typically low. Conversely, satisfying water demands for irrigation, meeting minimum-streamflow regulations and providing water to cities is typically far greater in the summer months.

All of this complicates the job of the water resource manager. As with any problem where the timing of supply is not well matched with the timing of demand, the best solution is effective storage. For example, if the annual floods of rivers like the Mississippi River could be captured and stored for use later in the year, drought would be less of a problem. In this way, sufficient water is often available to meet both ecological and economic needs provided the problem is considered regionally and over a long enough period of time.

The world is now in an era where neither the economy nor the environment can absorb much negative impact from the mismanagement of water resources. Regional watershed management may help shape policy to store water from the high-flow periods of watersheds and use that water to maintain flows for fish and wildlife requirements and meet the demands of the community during low-flow periods. Everyone should continue to strive for balanced policies that allow for the use of water without inordinate harm to either the environment or the economy. The need to balance the various and diverse interests that depend on water is more essential and more complex than ever.

WATER RESOURCES PLANNING

Most public utilities engage in some form of planning, although the extent and scope of planning vary greatly. Utility planning can be characterized by four general approaches: traditional supply planning, least-cost utility planning, integrated resource planning and total water management. Each iteration retains the tried-and-true methods of the preceding approach while expanding the scope of the planning horizon to include new issues and potential solutions.

TRADITIONAL SUPPLY PLANNING

Traditional planning for water utilities is not that different from traditional planning by electricity utilities, which can be characterized by its focus on utility ownership and control of all production resources (including centralstation power plants), its reliance on system and financial planning processes internal to the utility and its emphasis on the goals of minimizing electricity prices and maintaining a high level of system reliability. Risk avoidance also appears to be a paramount consideration in traditional utility planning. In the case of water, ownership and control of resources is more constrained (for example, by limits to groundwater withdrawals). However, the emphasis on utility ownership of the water delivery infrastructure (plants, pipes and so on) prevails much like the case of electricity distribution.

Shortcomings

Three principal concerns about the traditional approach have been advanced. First, forecast demand is taken as a given and virtually no attempt is made to integrate supply management and demand management options.

Second, the public-at-large, outside experts and government regulators generally have little or no involvement in traditional utility planning. Demand analysis and the assessment of supply alternatives takes place within the utility (or a single planning unit within the utility); only the final product is made available for review or regulatory approval. Often, major investment decisions are made with little or no oversight.

Third, traditional planning also tends to be confined to individual utilities in virtual isolation. The decision-making process excludes parties who not only have a vested interest but also unique insights about resource options. In sum, there is a tendency for traditional planning to be narrowly focused and exclusionary. Once a utility has committed itself to a plan, little room is left for altering the chosen course.

Supply-Driven Focus

Like other types of public utilities, the prevailing planning processes undertaken by water utilities have been internally driven and dominated by supply considerations.

The result has been an emphasis on maintaining reliable water supplies and, accordingly, the engineering of facilities for source development, treatment and storage and transmission and distribution of water. Water-supply planning generally takes the form of forecasting future demand and developing and analysing supply options to meet the projected demand level, plus a comfortable margin.

The result is a disaggregated planning approach focusing only on new supply alternatives, while initiatives in the areas of demand management and conservation are consigned to separate programmes. Moreover, conservation effects often are not reflected in revised demand and revenue forecasts.

Delivery, not Demand

Engineering considerations have always been central to water utility planning, at times to the exclusion of other perspectives. Engineering does take into account the benefits and costs of supply projects. However, the emphasis on supply options can result in additions to capacity that outpace growth in demand, which is a problem familiar to the electricity industry.Water's natural abundance in many areas may explain why the central water supply issue is that of engineering water delivery (getting water to where it is needed) rather than managing water demand. Historically, it was easy to presume that there always would be enough water to go around and that plentiful amounts would generally be so inexpensive that investments in demand management would not be cost-effective. The utility corporate culture, which understandably emphasizes selling more (not less) product, might also favour the supply side. At the very least, utilities traditionally have had little incentive to promote conservation.

LEAST-COST PLANNING

The shortcomings of traditional planning, along with significant external economic and political forces, gave rise to the current interest in least-cost planning, or integrated resource planning (IRP). Least-cost planning emerged in the context of the energy industries during the 1980s as a response to rising costs, a poor record of forecasting and growing concerns about environmental externalities. According to its proponents, planning allows regulators to be more proactive, that is, to actively affect utility decisions rather than simply react to them later on.

Least-cost planning emphasizes a balanced consideration of supply management and demand management options in identifying feasible least-cost alternatives for meeting future water needs. Compared with traditional planning, least-cost planning recognizes that water demand is malleable and that forecast demand does not have to simply be taken as a given in the planning process.Different definitions emphasize the minimization of rates, customer bills, utility revenue requirements and production (both capacity and operating) costs. The variety of available perspectives leaves open the question of whose costs are supposed to be least: existing customers, future customers, utility shareholders, or society at large.

Another unfortunate implication of the concept of least cost is that it suggests a single, optimal solution. Regulators and utility managers do not necessarily have identical views about the meaning of least-cost planning. According to one survey, "regulators tend to view least-cost planning with an emphasis on conservation, whereas utilities tend to regard least-cost planning as an integrated supply-and-demand analysis."

A Broader Context

Least-cost planning has come to be understood as the comprehensive evaluation of all supply and demand alternatives with the end result, in an attempt to minimize costs, of creating a flexible plan allowing for uncertainty and a changing economic environment. Cost minimization, resource diversity, risk management and flexibility are the hallmarks of least-cost planning.

In practical terms, least-cost planning can be characterized by the consideration of a diverse set of resource options (including conservation, load management, pricing and purchases from other producers); coordination among several departments within the utility; and the use of multiple resource selection goals, including those that address prices, costs, revenue requirements, utility financial condition, risk reduction, technological diversity, environmental quality and economic development.

Least-cost planning in no way abandons sound engineering practices, but it does place them in a wider context. It follows that least-cost planning can facilitate regulatory review of supply management and demand management practices and regulatory approval of capacity expansion projects prior to construction.

Least-cost planning also can be used to educate policymakers about the critical issues in utility supply decisions and induce utilities to aggressively engage in long-term strategic planning. Least-cost planning can focus on a particular utility or take the form of a statewide assessment of all utilities in a given utility sector, the latter of which is typically performed by a state resource management agency.

Demand Versus Supply

In contrast to traditional supply planning, leastcost planning gives much weight to the distinction between demand management and supply management activities, both of which can be used to achieve reliability goals. Least-cost planning recognizes that demand is malleable rather than simply a fixed input to planning models.

The demand side involves any strategy to eliminate or defer the need for an investment in new capacity by the utility, including load management, conservation and pricing strategies. The supply side involves determining the most efficient method of meeting growing demand, including investments in new capacity.

All demand management activities that decrease the demand for utility services tend to affect supply management since existing system capacity is released for other customers and other uses. That is, the freed or redirected utility capacity can be compared to that provided by more traditional means. Thus emerged the concept of "negawatts," meaning electricity "produced" through conservation and efficient use of existing electricity supply resources. A related idea is the trend to establish energy service companies, as compared with traditional electric utilities, who could market efficiency (or demand management) as well as electrical power.

Complexity, Risk and Cost

Least-cost utility and planning is complicated by the lack of familiarly with demand management, barriers to coordination with nearby utilities, the use of broad definitions of costs (such as those associated with externalities) and the inclusion of goals (such as resource protection) not directly attributable to the utility. Traditional planners tend to view least-cost planning as more risky, as least in terms of supply reliability. Advocates of least-cost planning would counter that planning actually reduces some forms of risk, such as the revenue recovery risk associated with excess capacity.

While highly pertinent to the debate, these disagreements can be overcome. One issue that merits especially careful attention, however, is that of the cost of planning itself. Certainly, the benefits of planning should outweigh the costs. This calculation itself poses a rational decision-making problem. The benefits of increased awareness and understanding and reduced ignorance and uncertainty can be substantial but are not easily quantified.

WATER ENVIRONMENTAL INFORMATION SYSTEM BASED ON GIS

Water is one of the most important resources of all humanity and plays a crucial role in the development of social economy and people's living. Now the situation of water pollution in China is becoming prominent and more concerned throughout the country. To keep sustainable development, rehabilitation of polluted water bodies is a task should be accomplished by great effort.

Water Environmental Management Information System (WEMIS) is established to achieve the collection, transmission, storage, maintenance and analysis of all sorts of water environmental information.

The construction of WEMIS base on new information technology can greatly upgrade the efficiency of environment management. Information involved in the water environmental management can be divided into three parts: water quality information, water pollution source information and related physical geography and socioeconomic information.

With the support of modern information technology such as GIS, WEMIS can rapidly employ the information well and truly. GIS and water environmental models are two analytical tools to establish a decision support system. It is necessary and potential to integrate these two tools in an appropriate way.

FUNCTION OF THE SYSTEM

Main functions of the system are as follows:

Maintenance of Water Environmental Data

It is the foundation of the system. Users will use the system to input, verify, modify, delete and backup the data involved in the water environmental management.

Support For Routine Operation

It includes enquiry of data, statistic of data, evaluation of water quality, evaluation of pollution sources, introduction to related physical geography, socioeconomic information and etc. Uses can comprehend the water environmental situation visually and quickly by all kinds of tables and graphs.

Decision Support For Management

It includes prediction of pollution sources, prediction of water quality, calculation of water environment capacity, allocation of pollution load and etc.

STRUCTURE OF THE SYSTEM

The system comprises six sub-system, and data flow diagram of the system is Figure below:

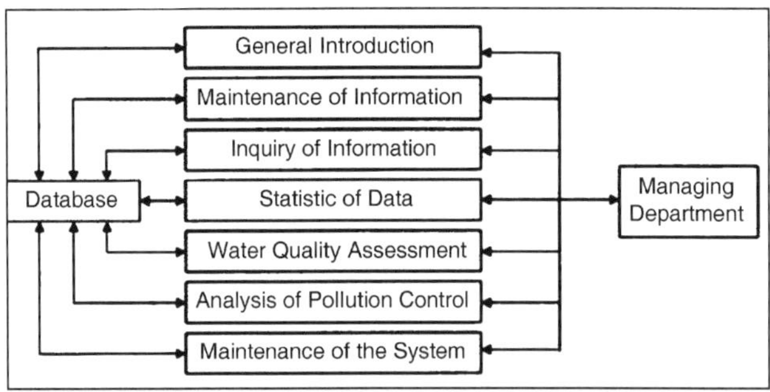

Fig. Data Flow Diagram of the System.

General Introduction

In this part, geography and socioeconomic situation, development of water quality and pollution sources, distribution of investigation station of a specific area can be comprehended in various formats such as text, table, picture and video etc.

Maintenance of Information

It includes input, modification and deletion of all the related data. In order to keep the integrality and consistency of the data, the system can provide validity verification.

Enquiry of Information

The system can form the corresponding Transact-SQL commands according to the demand of the user and send the request from the client application. Then the commands will be disposed by DBMS on server and return results to the client application.

Statistic of Data

Data statistic is an important working content in water environmental management. The system can conduct various statistics and generate report forms. Major items include monitoring station, water quality, and water pollution source.

Water Quality Assessment

The assessment of the current environmental situation will focus on two aspects. Firstly, the quality of water, *i.e.* pollution levels and distribution, will be addressed. Secondly, main polluting industries and municipal sewage discharges will be expressed using relevant parameters.

Analysis of Pollution Control

Data Mining is a main task of the system. The managers need to explore pollution-developing trend from the common transaction information and find necessary decision for pollution control. A series of water environmental models are involved in the system. These models include 1D and 2D water quality model, prediction model of polluting source and optimum method of polluting load allotment.

MAINTENANCE OF THE SYSTEM

In order to guarantee the security of the information of the system, all users will be assigned to a characteristic that determines the operations each user can perform.

INTEGRATION OF GIS WITH WATER ENVIRONMENTAL MODELS

GIS and water environmental models are two main analytical tools applied in the system to support decision-making. GIS based on common application provides measures for managing spatial information, but it can only act as system software.

For applying all functions of GIS to the water environmental information system, GIS should be integrated with application models in the corresponding field. One the other hand, water environmental models have distinct spatial character. Water environment situation is expressed by parameters indicated by monitoring cross-section and monitoring spot. The models can distinctly reveal the water environment quality in different area by introducing GIS, and indicate the distributing tendency of water environment quality.

APPLIED AREA OF GIS

GIS can help water environmental models in the aspect of data preprocess and data acquisition. It can also be used to establish postprocessor to output the result of water environmental models visually. For 1D water quality simulation, it can be efficiently applied to spatial display and analysis. With interpolation and contouring it can indicate the 2D water quality dispersion.

Spatial enquiry can be performed with GIS associated with result of water environmental models. For example, by using water quality model and line and polygon overly analysis technology in GIS, we can have comprehensive forecast and assessment over multiform pollution index of one water regime.

So integration of GIS with water environmental models is needed and potential. An ideal structure of integration should be adopted in the water environmental information system.

INTEGRATION METHOD

Different methods of integration differ from each other in two aspects: data interchange and user interface.

In the following discussion, four methods of integration will be analysed:

- Models and GIS are separate. They have user interface of each own, and exchange data with file. In this way the user interface won't have uniform style and the efficiency of data exchange is low.

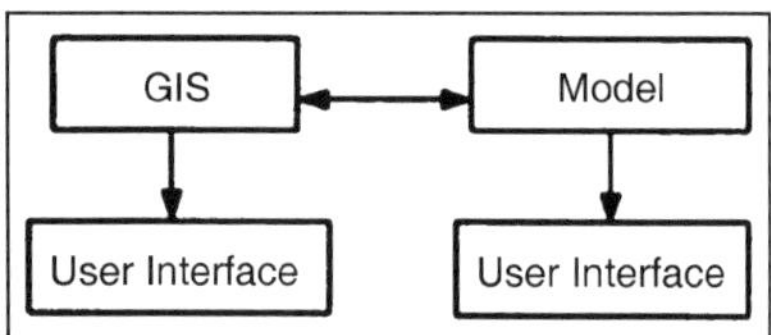

Fig. First Method of Integration.

- Models become an inline programme of GIS?*i.e.* Model is one module of GIS. Indeed some GIS products have analytical models can be applied in the water environment, but it is impossible expect one product can solve all the problems.

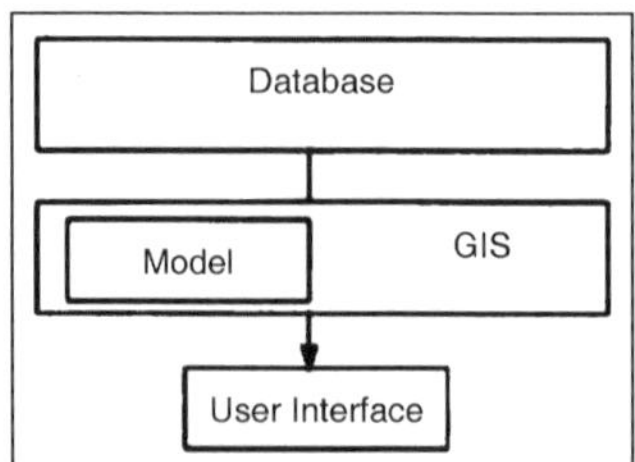

Fig. Second Method of Integration.

- GIS becomes an inline programme of model *i.e.* Models have some function of GIS. In the past, some researchers developed postprocessor to visualize the result of model. Although it improve the visual effect of the simulation, it fairly difficult to achieve the spatial analysis function of professional GIS product.

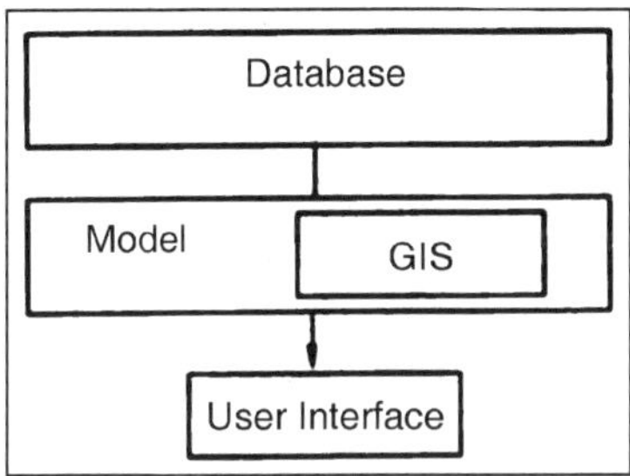

Fig. Third Method of Integration.

- GIS and model are separated application level of the system and have access to the data by information sharing. Users can operate GIS and model through a common interface.
  In this way, GIS and models can be integrated in a seamless method. Some calculating conditions of the model will come dirctly from database and spatial analytical function of GIS can be involved in the model. On the other hand, it becomes easier for the developers to establish a system that is more efficient.

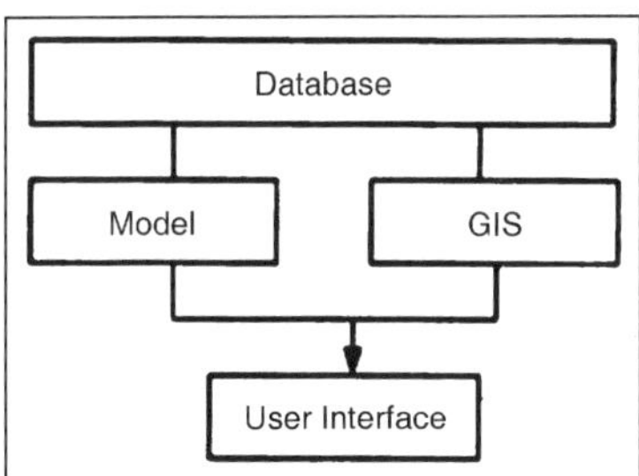

Fig. Fourth Method of Integration.

APPLICATION OF WEBGIS

In recent years Internet is rapidly spreading throughout the world. The Browser/Server (B/S) structure are more adopted by all kinds of information system. Contrasting to Client/Server (C/S), B/S has more flexibility and expandability and is easier to maintain, moreover, it can make the environmental information more open to the public.

But we aren't mean to substitute B/S structure for C/S structure. Instead, we will use both of them in the WEMIS. Information maintenance and analysis of pollution control will adopt C/S, and those functions that is more tied with information enquiry will adopt B/S.

WebGIS should be taken in the information system of water environment. Users will use Internet browser programme to send requests that can be accepted by Web server. The Web server startup CGI, ISAPI or NSAPI programmes and transfers the requests to GIS system. Accessing data by ODBC, GIS system will process the requests initiated by a client and return result. High-speed data transfer and high-performance network server should support WebGIS. But it won't be obstacle anymore in the near future with the fast development of modern network technology.

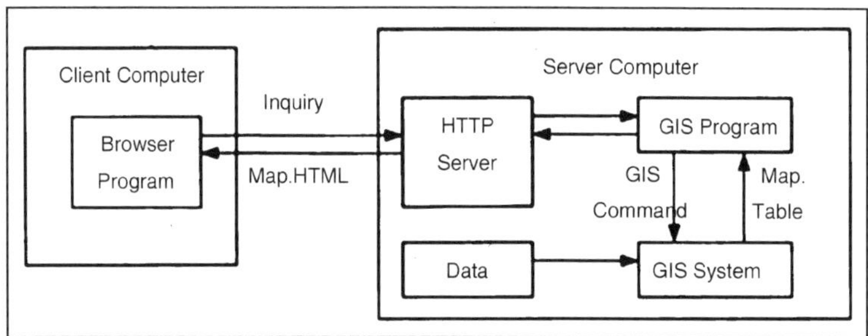

Fig. Enquiry Manner of WebGIS.

The development of WEMIS is an active and potential area. New technology such as Data Warehouse will be introduced into it to enhance the decision support capacity.

In the construction of WEMIS, GIS is an important technical element. The application of WebGIS is now at an elementary stage. There is a continuing need for more comprehensive research and practice on the application

INTEGRATED RESOURCE PLANNING

Integrated resource planning is a somewhat more encompassing term than least-cost utility planning, although the two are consistent and can be used interchangeably for many analytical purposes. In fact, the term "least-cost integrated resource planning" sometimes is used. The concept of integrated planning evolved in part to address the potential misconceptions and complexities arising from use of the term "least cost" as well as any unjustified bias against supply-side solutions.

COMPREHENSIVE AND PARTICIPATORY

Integrated resource planning encompasses the concept of least-cost planning, which emphasizes balancing supply and demand management considerations and identifying feasible planning alternatives that meet the test of least cost without unduly sacrificing other policy goals. Integrated resource planning is a more comprehensive evaluation system that goes further to emphasize the construction of various planning scenarios in which key variables and assumptions can be altered. These scenarios can be used to help utilities incorporate uncertainties, environmental externalities and community needs into decision-making.

Integrated resource planning also emphasizes the importance of establishing a more open and participatory decision-making process and coordinating the many water institutions that govern water resources. Thus, IRP encourages the development of new institutional roles in addition to new analytical tools. It also promotes consensus building and alternative dispute resolution over conflict and litigation. Importantly, planning does not preclude the development and use of markets or market-like mechanisms (such as competitive bidding), which in fact may be essential for the purpose of identifying least-cost opportunities.

Integrative Assessments

Like least-cost planning, IRP explicitly recognizes that demand management can be a cost-effective and viable resource option. In a somewhat broadened sense, IRP recognizes that demand management can help achieve multiple policy goals (such as cost control and pollution prevention). Advocates of IRP have long argued that better planning methods are needed to account for environmental and social externalities associated with expanding utility capacity. More recently, analysts have recognized that IRP can help utilities deal with uncertainties and risks as well.

The generalized concept of IRP also can be used to address short- and long-term community needs that span from environmental protection to economic development. Prescreening resource options and the construction of alternative planning scenarios can be used to evaluate the implications of a given resource mix on the utility, the environment and the community. In addition to integrating resource options, IRP also seeks integration along temporal (short- and long-term) and spatial (local and regional) dimensions.

Internal and External Linkages

Integrated resource planning clearly entails new roles and responsibilities for water-supply utilities. Integration means that environmental, engineering, public health, financial, rate-making, social and economic considerations all feed into the planning process. Planning data and information are linked internally to the other management activities of the water utility (physical facilities management, financial management, environmental management, research and development, economic development and public involvement).

Integrated planning also links water utility planning with external planning processes (planning by other water, wastewater and energy utilities; local and regional planning; river basin planning; and statewide, interstate and federal water planning and policy). Some of these relationships are formal (as in permit processes involving state water resource or drinking-water regulators), while others are less so (as in the use of regional water planning data by the utility to develop forecasts). For the water sector, a particularly important issue is the relationship of water utility planning to the activities of various government agencies whose policies may constrain utility planning choices. Clearly, better coordination mechanisms could be established among the three or more state agencies that regulate water systems from the public health, natural resource and economic perspectives. If IRP does nothing else but facilitate coordination and improvement in the institutional structures governing water resources, it will be worth the effort.

In the 1990s, total water management emerged as a potentially salient concept for water and wastewater utilities. Total water management reflects the philosophy that water resources should be managed for the greatest good of people and the environment with opportunities for participation in water policy by all segments of society.

While not strictly a planning model, total water management seems to encompass the basic principles of IRP. It also has considerable symbolic connotations. According to a white paper by the American Water Works Association, "Total water management recognizes the paradigm shift from considering water available in unlimited quantities to understanding water supply as a limited resource." Total water management seeks to inspire the water industry to embrace such ideas as sustainability, stewardship, unified water resource policies, watershed and ecosystem management, water conservation and the importance of public and political support for water management decisions. Total water management recognizes that water resources are a part of numerous complex systems, both natural and social. Advocates of integrates resources planning (encompassing, for example, water, energy and land-use planning) make a similar point. These perspectives present numerous intellectual, analytical and evaluative challenges.

The term "total water management" and its manifestations are more normative and prescriptive than the seemingly value-neutral realm of rational analysis. The explicit consideration of values in planning can be uncomfortable, particularly for members of the many science-based disciplines involved in water resource issues.

But in making policy choices, trade-offs among competing values are inevitable. A paradigm that allows for a dialogue about how to make these trade-offs should be welcomed by the water industry.

PLANNING IN PRACTICE

The practice of planning requires attention to the potential barriers to success. Three key areas of concern are:

- Access to analytical tools and adequate information,
- The level of commitment of utilities and regulators to considering and pursuing new options,

- The consistency of approaches and methods within the real-world context of existing utility and regulatory practices.

If these issues are relevant to least-cost planning for energy utilities, they are as much or more applicable to the case of water.

Information Base

For water systems, information resources vary substantially and the need to develop data processing and analytical capabilities is clear. Attention should be paid to the design and implementation of planning strategies that minimize the effects of inadequate information.

Participants

The second barrier involves the attitudes and dispositions of those involved in resource planning. In part, success in IRP will depend on whether the prevailing corporate culture accepts conservation and related concepts as legitimate utility goals.

Realistic Approaches

Finally, successful IRP requires good practice as well as good theory. Water utility managers need practical methods for dealing with new forms of uncertainty, such as the revenue uncertainty associated with conservation.

SIMPLE ANALYSIS ON FUTURE WATER RESOURCE

Recently many researchers have paid attention for water resource problems and many people are worrying about water conflict in the future. World water forum indicates probability of water shortage by increase of irrigated area with population increase and said right of secure water use 40 litre per day for all people in all over the world as target. However the achievement of this target is not only difficult but also lack of water happens everywhere.

Such problems are complicated in an international river and international organizations are needed for management of the river. The Mekong River is going through six countries and the river has 800,000km2 as the area with 4,200km length. Lack of water is worried in the future because Lao and Cambodia are developing for irrigation and hydro power plant.

Especially Vietnam makes caution for such water resource developments. The design of development in the region needs holistic and comprehensive planning and the evaluation and the assessments. Especially, multi– directional and multi-scale discussions are necessary for development in the headstream of the river.

MRC (Mekong River Commission) compiles database consisted of hydrologic and meteorological data in lower four countries. Also remote sensing and digital map data are available and useful supported by GIS. The ultimate target is discussion on the development design by multi-direction and -scale organically using those data. In this paper, simple analysis of water balance and drought problems are mentioned.

DATASETS

Satellite data of AVHRR/NOAA is used for grasping vegetation condition and land use. The data is received and complied in 1998 by ACRoS (Asian Centre of Remote Sensing) in AIT(Asian Institute of Technology).

Meteorological and hydrological data are obtained from MRC Year book in 1993. Elevation data is obtained from GTOP30 of USGS and extracted as watershed area. Land use data is depending on IDI (Infrastructure Development Institute, Japan). Study basin is considered only lower area from Chanseing.

DROUGHT PROBLEMS WITH LAND USE CHANGE

Study Basin

Virtual area was introduced in this study for simple consideration without complex land use and reservoir management. The virtual area is given as unit area with only three land uses, which are forest, paddy and urban area. Considering the Mekong basin, upper and lower regions are covered with forest and paddy field, respectively. Urban area is less than 1 per cent. It was supposed that urban area is developing in the lower region. Land use data in 1997 show us 17 per cent crops and paddy fields.

Other areas are forest and glass lands. According to development, crops area will expand from flatter area. Potential area for crops is obtained from slope calculated elevation data. Ratio of slope less than 20m/km to whole area is 58 per cent. Now supposed rate of land use change on time is supposed. Crops field linearly increases until potential area and simultaneously forest area decreases. Urban area increases in crops field until 10 per cent in the same time range.

Hydrological Processes

Precipitation data at Pakse is used as input to study basin and is distributed uniformly in a whole basin. It is supposed that evapotranspiration in crops field was estimated as potential evapotranspiration 120mm/month in period of water supply. Kazama etc. estimated monthly evapotranspiration from forest in Thailand and Vietnam.

Those values are used. In urban area, only interceptional evapotranspiration is considered as 10mm/month, where monthly rainfall is equaled to urban evapotranspiration if the rainfall is less than 10mm/month.

Although crops field area is grater than paddy field now, it is supposed that paddy field will dominate crops field by irrigation system expansion. The storage height of paddy field is 100mm/month from May to September, rainy season. When the result of water balance is not enough for the storage, the lack will be charged by run-off after June. Storage period is from May to September and all storage water discharge in October. Storage in urban area is ignored. In forest area, run-off and storage is expressed by the Tank model.

Run-off ratio to net rainfall is 0.9 and infiltration ratio coefficient to storage height is 0.02. Concerning infiltration in paddy field, the rates are 2mm during storage period and 2 per cent to net rainfall during unstorage period, respectively.

When run-off is negative by water balance, paddy field releases storage water to river, which has 5mm/month as minimum discharge. After this the supplement for paddy field is charged from run-off of next month. These conditions are shown in Table.

Table. Input Hydrological Amount for Land Uses

	Forest	**Paddy**	**Urban**
Rainfall P	Constant: Pakse :MRC Hydrological Year Book, 1993. Tokyo, Science record book		
Runoff R	Calculation by Tank	Water balance	
Infiltration I		Dry: 2mm Wet: 0mm	0mm
Storage S		Full: 100mm Not Full: 0mm	0mm
EvapotranspirationE	Table 2	Full: 120mm Not Full: Forest.	Interception: 10mm

Table. Evapotranspiration (mm)

	Jan.	**Feb.**	**Mar.**	**Apr.**	**May**	**Jun.**	**Jul.**	**Aug**	**Sep.**	**Oct.**	**Nov.**	**Dec.**
Paakse	10	10	20	30	100	130	140	150	150	100	10	10
Tokyo	10	10	30	60	80	100	120	120	90	50	30	10

Estimation of Drought by Water Balance

Water balance is calculated by the following equation.

$$R = P + Q - E - I - DS$$

Where R is run-off, P is precipitation, Q is discharge, I is infiltration and deltaS is storage. Every amount is used as mentioned. Run-off in the study basin is estimated from water balance considering the land use change in time.

Water balance in cases of time 0 and 80. Here, time is conceptual under virtual scale. Although there is a little bit of difference, evapotanspiration in summer and run-off in June and July are decreasing in time 80 with deforestation. The reason is that storage water is increasing as crops area is increasing. In summer period, evapotranspiration in forest also is more and run-off is increasing, on the other hand, less run-off produce. Lack of water happens during planting season and it will influence to the month of May after the expansion of crops area. To compare with these results, the calculation of precipitation and evapotranspiration in Tokyo in the same land use as the Mekong. This will explain the importance of availability in water resource in planting season on the scenario, which is expansion of paddy field. In Japan, snowmelt water can supply water for paddy field in this season, however the Mekong region has drought. The existence of dry season makes chronic drought in the Mekong. Water demand is also increased not only by expansion of crops but also urban area. Urban water demand height is calculated as follows.

$$Wd = U_w \times Pd$$

Where Wd is household water height, Uw is urban demand, $20m^3$/month and Pd is urban population density, 6,000people/km^2. Urban water demand is not variable in Japan in recent years, so that, the value in Japan is useful. Urban population density is referred from Singapore and Hongkong. As the result, Wd is 120mm/month. Drought happens when water demand excesses run-off from the basin. This means R<Wd at lowest point. The number of drought months by land use change is indicated. This shows Tokyo's phenomena. In the Mekong region, the number of drought months increases as crops area increases. The reason is why base flow from forest decreases and storage amount increases. In time 0, drought happens almost once a month, which is season of plantation. Widening paddy field extends drought months after May. In the case of Tokyo, decrease of drought months is shown due to decrease of evapotranspiration in urban area. These results have some assumptions and difference from actual phenomena.

RELATIONSHIP BETWEEN POPULATION AND IRRIGATION AREA

The importance in water resource management is practice of water supply for everyone as mentioned in introduction. Therefore, let us consider potential irrigation area to distribute 40l/month to every person as target of World water Forum. Monthly water demand Dw is,

$$Dw = Pd \times Dw_p$$

where Pd is population density and Dwp is personal water demand with 40l. Population density 85 people/km^2 is estimated from 67.4million people and 793.1 thousand km^2 in the Mekong river basin. Using these values, Dw is 0.102mm/

month. In February, which is the most drought month, run-off from study basin is 15mm/month obtained by water balance. In this case, evapotranspiration in forest is equal to crops field. If 100mm of water is necessary for irrigation in dry season, irrigated area Ai is,

$$Ai = (R-Dw)/100$$

where R is 15mm and Dw is used as above results. The variation of potential irrigated area is if population will be double. The Mekong river basin shows a little chance of population increase. Almost current crops field can supply to double population, however, little expansion of irrigation area rapidly decreases water supply water for one person. In this result, 15 per cent irrigation area to a whole basin is maximum and the expansion more than this needs more efficient utilization for water. Expansion of irrigation area is national policy in Laos and Cambodia and it will be apprehensive about water supply in the future.

INDIGENOUS SYSTEMS OF WATER MANAGEMENT

ndigenous knowledge in India has always developed practical ways for society to live in a sustainable manner with nature, in full respect with the diversity of agro-ecological climatic zones. There are various types of methods of water harvesting in India, characterized by: use of local resources and technology; community-based operation; community-driven decentralized water management; and sustainable conservation and use of natural resources. The author notes the need for revival of systems using indigenous knowledge, including interventions to understand traditional systems and use of indigenous knowledge; mobilization of community around land, water and forest; participation in rejuvenating old structures and construction of new structures; and creation of new village level and river basin institutions.

Large-scale water projects which work against nature's water economy and people's sustenance requirements have been designed by destroying water technologies which were ecologically more sustainable and socially more just. The tank system of South India is among the indigenous alternatives which has survived over centuries. This system consists of a single series of several hundred and in some cases over a thousand reservoirs linked together and forming such continuous chains of works that not a single drop of water falling in the catchment is lost in times of drought, and very little is lost during normal periods. Major Sankey, one of the first engineers of Mysore State, who concentrated on the systematic repairs of tanks, stated that 'to such an extent has the principle of storage been followed that it would require some ingenuity to discover a site within this great area for a new tank'.

These tanks play a central role in irrigation even today. In the Ravalseema region. in the southern part of the Krishna basin, tanks irrigate 620,000 acres while major and minor irrigation projects cover 427,000 acres. In the Anantapur region, river water was diverted with the help of sand dams. These sand dams were constructed in areas like Koppalakonda, Penakacharala, Kalluru, Tarimala and Rachepalli of Anantapur taluk, Panidi, Appeyipeta, Naganapuram and Chitrachedu of Gooty taluk. In places like Ramapuram, Kallavapalle-and Budigamma, the surface flows in the Pennar were diverted through masonry dams called 'Panthams'. In Hospet, Hagari, Rayadurge and Kudligi taluks, channels were constructed to draw water from rivers for irrigation. Similar types of irrigation schemes have also been reported in Dharwar, Bijapur, Sholapur, Satara, Sangli and Ahmednagar districts. Two such major schemes were irrigating nearly 580 acres in Bijapur district. In Madras Presidency, irrigation through small tanks and canals, which the villagers managed themselves, collectively irrigated an area equal to that irrigated by all the larger works which have been constructed by the British government in that Presidency.

An example of indigenous water use was the widespread system of 'Ahars' and 'Pyres' used for irrigation of paddy fields in South Bihar as reported by Sengupta.4 An 'Ahar' is constructed by erecting an embankment 1 or 2 metres in height on the lower ground. From the two extremes of this embankment two other embankments are constructed so as to project towards the higher ground, gradually diminishing in height as the ground level rises and ultimately ending at the ground level. 'Ahars' were built on drainage rivulets to collect water. Constructions with sides more than a kilometre in length and irrigating over a thousand acres of land were frequently found. 'Pyres' on the other hand were systems devised for utilising the water which flows through hilly rivers running from south to north and intersecting the whole country. 'Pyres' were laid off from the rivers to carry water to-agricultural fields. Some of the largest 'Pyres' were 20 to 30 km in length, fielding a number of distributaries and irrigating may be 100 villages.

The system protected agriculture to such an extent that when there was a famine in other parts of the country, Gaya district, where 'Ahars' and 'Pyres' were most evolved, was not affected. However, this immunity was eroded following the deterioration of these irrigation works, which was primarily a consequence of breakdown of economic and social systems which ensured maintenance of the water systems.

The irrigation works in pre-British India were managed by a variety of social organizations within the village community. Usually, the structure of such organisations included a collective of all the beneficiaries of irrigation works, and was headed by a leader. Based on the leadership, the system was known by different names in various regions.

In Maharashtra districts of the basin, the bandharas were managed by such water committees. The task of these committees was to maintain the earthern dam or diversion from the river and to desilt the canals. Similar work was performed by committees in Bijapur, Dharwar, Raichur and Bellary districts of Karnataka. Similar committees were also found in the districts of Andhra Pradesh where there were known as 'pinnapeddandarule' or 'peddandarule' system. The difference between the two was in terms of the members who constituted them. In the 'pinnapeddandarule' system, youth were preferred as the desilting of canals involved hard physical labour. In Krishna district, less labour intensive work was involved hence the membership rules were flexible. Work involved in desilting or erecting canal banks or rebuilding canals was equally shared by all the beneficiaries. Each member or beneficiary was supposed to do his share of work in proportion to the land held. The committee levied a fine on any one who failed to do his share of work, irrespective of the size of his landholding.

Similarly in South Bihar, both the construction and maintenance of water systems was collectively managed. Each cultivator had to contribute labour to a collective system called 'goam' to repair embankments and desilt channels. In South India this practice was known as 'kudimarammath'.

Allocation of water within the villages was also managed by the cultivators themselves. A system known as 'parabandi' was in operation which regulated the distribution of water among the villages from a common source to ensure fair distribution to all the villages. In case of some large works, the rights of each village were formally recorded. In others, the regulations were largely customary. If conflicts arose, they were resolved according to local rights and regulations.

When the British arrived in India, they had no expertise in water management, since agriculture in Europe has never been dependent on irrigation. Arthus Cotton, the founder of modern irrigation programmes wrote: There are multitudes of old native works in various parts of India. These are noble works, and show both boldness and engineering layout. They have stood for hundreds of years....

When I first arrived in India, the contempt with which the natives firstly spoke of us on account of this neglect of material improvements was very striking, they used to say we were a kind of civilized savages, wonderfully expert about fighting, but so inferior to their great men that we would not even keep in repair the works they have constructed much less even imitate them in extending the system

Greenway, while writing on farming in India, commented on the lack of advanced engineering knowledge of the British which had led to the catastrophe in Sheffield because of the Bradfield dam. A comparison naturally presents itself between the dam of the Bradfield reservoir, which failed, and the Indian model which has been so long and in so many instances successful, and which if rightly constructed and faithfully attended to, may be regarded as ensuring the maximum of efficiency and safety.

Empirical as the process may appear, practice has made it perfect. Engaged in a continual struggle with the powers of nature, contending with volumes of water far larger, floods far heavier, tempests more violent, than any known in England, the Indian engineers have been forced into devising means, not only to enable a bank to stand a given presence of water, but also to provide resources against contingent risks and accidents, which latter provision, strange to say, appears from the evidence on the inquest of Sheffield not invariably to be considered a part of the engineer's duty.'

The British, however, had to control river water in India keeping in view their economic interest in increasing revenue. In Rajasthan they controlled water to maximise the revenues derived from their monopoly in salt trade, in Bengal to protect their transport network and increase revenues from agriculture. Water politics was thus as transient as the resource itself, to create obey areas and command areas, *i.e.*, areas and people inhabiting them which are callously devastated to cater to the needs of areas and people which are industrialised, urbanised, and politically and economically powerful.

Langdon Winner has suggested that politics is embedded in artifacts. Large water projects definitely have politics built into them. In an attempt to control rivers, they also control the lives of those who depend on the river and its basin for livelihood. Worster has proposed the thesis that making more demands on the earth and devising the means to fulfil them leads to an unequal power division in society. 'Intensification of use eventually must give rise to potent anti-democratic forces whatever their guise may. Karl Wittfogel's idea of a hydraulic society captures the reality of the social organisation created by large dams. Wittfogel's theory is one of power, he holds that control over water leads to control over people.

Wittfogel, however, like Marx before him, perceived the hydraulic society as a reality in Asian civilizations over thousands of years. They assumed that the pervasiveness of and decentralized network irrigation systems was linked to centralized power and that specific individuals conquering river waters turned into a power elite. What Marx and Wittfogel failed to perceive from a distance was that irrigation systems in India were managed by cooperating communities, not dominating bureaucracies, and that decentralized maintenance and use rather than centralised control was the characteristic of India's ancient irrigation systems.

Sengupta has challenged Wittfogel's thesis that domination was characteristic of Asia's water systems. Vast networks of irrigation systems are not necessarily large projects. They can be a closely knit network of micro-projects, each managed locally in terms of construction, maintenance, allocation and conflict resolution among users. Further, stagnation has not been a characteristic of these traditional irrigation systems. Instead, flexibility was often displayed.

The cropping patterns were changed annually according to the availability of irrigation water in a particular year. With water resources under local control, local decisions on land use involved less risks and higher certainty. On the other hand modern canal irrigation from large dams centralises water control and distribution. The time of water supply rarely corresponds to the time of requirement and certainty is far less than it would have been had the cultivators themselves distributed the same amount of water through local resources. Correspondingly, they have much less scope to alter their cropping and irrigation practices to suit the availability of water.

Centralised societal domination is, therefore, linked to centralised water control not to vast networks of decentralised waterworks. The power of the modern state over people as exemplified through large dams is qualitatively different from the social organisation of indigenous irrigation systems. Such political control does not merely violate human rights in the present, it threatens to deny future generations the right to life-support systems. Large water projects interfere in a major way with the natural flow of water and the hydrological cycle. Ecological hazards are intrinsically associated with them, which in turn generate another level of conflicts over natural resources. Social control over water use in indigenous systems had prevented both over-use and abuse of water, and avoided a conflict between the use of water for human consumption and its functions in the maintenance of essential ecological processes, central to which is the water cycle.

CLIMATE RISK AND AGRICULTURAL WATER MANAGEMENT

African countries and parts of India lack public or private infrastructure to provide storage to mitigate the variability of rainfall. Increasing the control of water resources is a direct method for managing the risk associated with climate variability and water storage is the most common approach to increasing water control.

Storage options range from bunding and on farm impoundments to retain run-off from individual storms for supplemental irrigation, to large scale dams that can retain high flows, reducing the damage from floods and also providing water during dry seasons or droughts. In general, wealthy nations have invested in large scale storage infrastructure to achieve reliable water supply for domestic use and agriculture and for protection from extreme climate events, such as floods and drought.

An analysis of regional differences in the volume of storage relative to need indicates that the tropics generally, and the Caribbean, Africa and Asia specifically, have low levels of storage relative to need. It is apparent that the nations with the lowest per capita GDP have the greatest needs for additional storage. On average, less developed countries face greater climate variability and the resulting accumulated economic losses hinder their ability to invest in the storage needed to mitigate the variability effects.

STORAGE AND RELIABILITY

The degree of water control that storage provides is a function of the size of the storage, the variability of the water flows and the magnitude of water demand. Large scale storage captures run-off from a distributed spatial area over which the heterogeneity of rainfall in space and time is smoothed. As a result, it is more reliable than storage that captures run-off from smaller spatial areas.

Depending on the size storage volume, large scale storage may also accommodate a second growing season, provide protection from floods and generate hydroelectricity. However, environmental and often social costs often increase with storage volume, due to submerged areas that displace people or habitat. Historically, large scale irrigation schemes in sub-Saharan Africa have not paid off economically with low rates of return on investment.

They have been plagued by high per hectare costs, low profitability, lack of transparency and lack of farmer involvement. A study of 42 irrigation projects in sub- Saharan Africa found that between 1970 and 1984 projects were plagued by low rates of return (<10 per cent) and high per hectare costs (averaging $24,500/ha), while projects implemented after 1984 showed some improvement. Successful projects had per-hectare costs of $3000 - $5000 and many were rehabilitation projects. Preliminary research results from a study of the economic potential of irrigation investments indicates that major future investments in large scale irrigation are unlikely for economic reasons even considering "add-ons" to existing dams or those to be built for hydroelectricity and flood control.

In contrast, small scale and community managed irrigation have been generally more successful, although most existing studies were completed at the pilot scale only. A variety of studies have examined the economic viability of various approaches at local scales but there are few studies of there reliability in response to climate variability at a planning perspective.

Smaller scale storage offers the benefit of more local control and less externalities in terms of submerged area. Pilot studies of surface impoundments (farm ponds) in Kenya and Burkina Faso have found significant potential increases in yield and productivity through supplemental irrigation and extending the growing season. Improved water control may also be achieved through methods that focus on the control of evaporation, such as conservation farming, drip irrigation, furrowing and leveling of fields.

These methods have also tended to be the most economical, although there exists low rates of adoption for reasons that are not entirely understood. In the Kenya study, one factor was the high rate of water loss due to seepage and evaporation. Cost estimates for these approaches are wide ranging and overlap with the estimates of the most economical large scale projects.

In a forthcoming report, IFPRI estimated the cost of traditional methods such as water harvesting at $600 to $1,000/ha, individual irrigation systems such as pumps and distribution lines at $1,500 to $3,000/ha and community level irrigation, including small dams, at $3,000 to $8,000/ha.

RESIDUAL CLIMATE RISK FOR AGRICULTURAL WATER MANAGEMENT

The reliability of water supply is a key factor in the success of interventions to improve agricultural productivity. Since water supply is less reliable at local scales, and water management investments are most likely to be made in local scale approaches, due to the economic reasons cited above, consideration of the climate risk that is not covered by these approaches is warranted. We label this remaining risk "residual climate risk."

Increasing the control of water resources available for agriculture reduces vulnerability to climate variability and leads to greater agricultural productivity. Water management strategies differ in the range of climate variability they are able to control. Local water management approaches provide mitigation of slight or moderate departures from normal rainfall, but are limited in mitigating more severe anomalies, such as droughts or extended dry spells.

Because the drainage area over which run-off is collected is smaller, the reliability would on average be reduced. Rainfed agricultural systems, which represent 93 per cent of SSA agriculture, are highly vulnerable to climate variability. Soil management approaches, such as conservation farming, may improve productivity in near normal and below normal rainfall years. In Kenya, tied ridges increased yields on average, but are economically viable only in seasons with moderately low rainfall, as yield increases do not cover the added setup costs when rainfall is either very high or very low. Thus, such approaches leave residual climate risks including dry spells during critical growing periods, droughts, and lack of protection from floods.

Farm scale irrigation systems, such as groundwater irrigation and on-farm surface water collection, and community-managed systems, reduce the vulnerability of farmers to dry spells and intra-seasonal rainfall variability through supplemental irrigation and may allow extension of the growing season. A pilot scale study of farm ponds in Kenya and Burkina Faso found them to be profitable, but included a limited range of rainfall variability during the three year study.

A similar study in Kenya determined that losses from evaporation and seepage limited their effectiveness and may have contributed to the low rate of adoption there. Because the catchment area of on farm water collection systems is relatively small, they are less able to mitigate the spatial heterogeneity of rainfall and thus water supplies are less reliable than larger scale efforts. Residual climate risks include drought and floods.

Groundwater irrigation may provide protection from drought and dry season farming, depending on the depth to water and the volume of the aquifer, and its potential application within Africa is not well known. Experience from India makes clear that groundwater can be a boon to agricultural production but also is vulnerable to over-exploitation.

Large-scale public irrigation system are surface water systems typically with large storage volumes collecting water from sizeable contributing areas and delivery systems covering a large area. These systems are designed to provide resilience to most droughts and to allow multiple cropping seasons and provide the highest potential protection against climate risks. Still, actual reliability of water supply for an individual farmer depends on one's location and priority within the system. For example, reservoir systems where water is shared with drinking water supply systems and hydroelectricity production often place the lowest priority on agricultural water and the reliability of agricultural water deliveries is compromised.

Because the cost of dam construction is so large, rccent studies indicate that irrigation expansion via dams will only be economical where the primary water use is hydroelectricity generation. The residual climate risks of these systems then include extreme or multiyear droughts and competition for water resources with hydroelectricity and drinking water during times of scarcity.

CLIMATE, WATER, AGRICULTURE AND DEVELOPMENT

Exposure to a high degree of climate risk is a characteristic feature of rainfed agriculture in the drylandsi of sub-Saharan Africa and parts of South Asia, which were largely bypassed by the Green Revolution, and where poverty and food insecurity remain most prevalent. Agriculture represents over 90 per cent of water withdrawals in India and sub-Saharan Africa (SSA). In SSA, 93 per cent of agriculture is rainfed, representing 70 per cent of the population's employment and 35 per cent of GDP.

SSA also has the least-developed water storage infrastructure needed to manage the variability of rainfall. As a result, economic development in SSA remains particularly vulnerable to the vagaries of rainfall. While climate change may increase the challenge, already present climate variability is a major impediment to economic growth in SSA as a result of the large fraction of economies that agriculture represents and its vulnerability to climate anomalies.

Of the 183 million hectares of agricultural land in SSA, only about 9 million is under some form of water management.

While investment in expanding irrigation seems an auspicious way to improve agricultural productivity, estimates place the area of additional irrigation that would be profitable investments as part of dam-based schemes at only 3 million hectares.

For economic reasons then, small scale water management and storage approaches are likely to be a key component of efforts to increase agricultural productivity. The same preliminary IFPRI report estimates that small scale approaches might be profitable investments on an additional 38.2 million hectares. However, smaller-scale water management systems are best prospect for improving productivity under near-normal or moderately belownormal rainfall conditions. They are much less capable of managing climate extremes, such as floods and droughts.

Farmers will continue to face considerable climate risk, and extreme events can reverse development gains made over many years. A single drought or flood could set back all the agricultural development progress that result from improved local water management. We term this remaining climate risk, "residual risk," and recommend that a strategy for investing in agricultural water management should include a multi-pronged approach to dealing with the full range of climate variability, including not only moderate years but also the extremes.

CLIMATE VARIABILITY IS AN OBSTACLE TO AGRICULTURAL DEVELOPMENT

A growing body of evidence links unmitigated hydroclimatic variability to poor economic growth in developing countries. Cross country analysis show that in most poor countries, climate variability is high, infrastructure in lacking and GDP is correlated with rainfall.

In a study of climate impacts on economic growth of the countries of sub-Saharan Africa, drought was found to be the dominant climate risk, having a significant negative effect on GDP growth in one third of the countries. Droughts are also the world's most expensive disaster, destroying the economic livelihood and food source for those dependent on the agricultural sector or their own food production.

The World Bank estimates that economic growth in Ethiopia is reduced by one third due to hydrologic variability. A single drought over a 12-year period is estimated to reduce economic growth during the whole period by 10 per cent. Drought impacts in Kenya associated with the La Nina of 1998 to 2000 resulted in losses totaling 16 per cent of GDP. Floods destroy infrastructure, disrupt transportation and economic flows of goods and services and can lead to contaminated water supplies and the outbreak of waterborne disease epidemics, such as cholera.

In the Mozambique floods of 2000, over 2 million people were affected and damages were estimated at 20 per cent of GDP. Flood damages in Kenya associated with the El Nino event of 1997-1998 were estimated at 11 per cent of GDP. The effect of these hydrologic extremes can be devastating in any country, but especially in those with enhanced vulnerability due to high dependence on agriculture and deficient infrastructure. This is the case throughout SSA.

Climate exerts a profound influence on the lives of poor rural populations who depend on agriculture for livelihood and sustenance, who are unprotected against climate-related diseases, who lack secure access to water and food, and who are vulnerable to hydrometeorological hazard. Climate variability is arguably the dominant source of consumption risk in smallholder rainfed agriculture in the dryer environ-ments of much of sub-Saharan Africa and India.

Within farming communities, because the relatively poor have less capacity to buffer against climate risk through own assets or financial markets, they tend to experience disproportionate livelihood risk in the face of climate variations. Current understanding of the mechanisms by which climate variability impedes agricultural development, and hence efforts to ensure food security and reduce rural poverty, provides insights into opportunities for intervention.

The various mechanisms by which climate risk impacts households combine with other factors to trap rural populations in chronic poverty. A dynamic poverty traps occurs when there is a critical threshold of household assets, below which individuals are unable to accumulate the necessary resources to escape poverty.

The tendency for risk tolerance to decrease with decreasing resource endowment contributes to the higher opportunity cost of climate risk for the relatively poor, and hence the locally-increasing marginal productivity that contributes to the existence of the multiple equilibria associated with a poverty trap. Ex-post coping response to severe or repeated climate shocks can push households to divest productive assets to a point below the poverty trap threshold.

Climate risk also impacts institutions in a manner that further constrains economic opportunities and hence reinforce poverty traps at the household level. Examples include the increasing cost of food crisis relief competing with agricultural development for shrinking donor resources, and the widespread reluctance of lenders to serve smallholder rainfed farmers.

While much is known about how climate risk impacts agriculture, less is known about the magnitude of the impacts or the magnitude of the livelihood benefits of feasible opportunities for managing climate risk. Despite the known impacts of current climate risk and growing concern about future climate change, climate risk management remains conspicuously absent from many analyses and regional development strategies.

We speculate that this is because the agricultural community has long regarded climate as part of the environmental baseline and not a resource with options for management. Development strategies that recognize climate risk generally limit intervention to expanded irrigation or improved water management. The Comprehensive Africa Agriculture Development Programme (CAADP), endorsed by the African Union, cites "vagaries of climate and consequent risk that deters investment" as one of six key challenges to achieving a productive and profitable agricultural sector across Africa.

Its strategy for addressing this constraint emphasizes "extending the area under sustainable land management and reliable water control systems". However, recent assessments and strategies, such as the Comprehensive Assessment of Water Management in Agriculture, increasingly recognize that feasible investments in water management for the drylands of SSA and parts of South Asia are necessary but not sufficient for overcoming the challenge of climate risk to development.

The infeasibility of achieving a high level of water control across the vast dryland farming regions of Africa in the near to medium term, and increasing stress on groundwater and surface water resources in much of India point to the need to exploit every opportunity to deal with the residual climate risk that water control systems cannot mitigate.

CLIMATE CHANGE WILL INTENSIFY AGRICULTURAL WATER CHALLENGES

Although the effects of climate change from anthropogenic forcing on the use of water resources in the world remain difficult to project, anticipated climate change combined with other drivers of change is likely to intensify current agricultural water management challenges in Africa and India.

The effects of population growth and increasing water demand, which are often but not always coupled, are likely to be a more significant source of water stress than climate change when considering changes to mean precipitation and run-off. Increasing temperatures in all regions are expected to increase evaporative demand, which would tend to increase the amount of water required to achieve a given level of plant production if crop phenology and management are held constant.

However if cultivars and planting dates were to remain unchanged, accelerated crop development in response to temperature increases would tend to have the opposite effect on water requirements. Increased temperatures are also expected to increase evaporative losses of surface water resources. The magnitude and even direction of projected changes in precipitation are quite uncertain. Based on how many of the 21 climate models used in the 4th IPCC assessment (AR4) predict increases vs. decreases in annual and seasonal rainfall across Africa and Asia), decreased annual rainfall is very likely in Mediterranean North Africa and likely in much of Southern Africa particularly for the southern winter. Increases in rainfall are likely for much of Eastern

Africa particularly in the northern winter, and for the summer monsoon over much of peninsular and eastern India.

These climate models are divided between wetting and drying trends in most of West Africa and western India. The two models within this set that best captured the extended dry period (1970s-1990s) in the West African Sahel predict opposite response to projected greenhouse gas forcing.

There is, however, a consensus that climate change will tend to increase the variability of rainfall and decrease the natural storage provided by snowpack and glaciers, such as in the Himalaya that feed the ricewheat belt of the Indo-Gangetic Plains. The concern pertaining to climate change impacts should not overshadow the present challenges that climate variability poses to agricultural development.

While the future impacts of climate change remain uncertain, climate variability persists as a challenge to development and as an impediment to meeting the Millennium Development Goals. Over the next decades, while climate change trends may begin to have some effect, droughts and floods associated with climate variability will continue to ravage vulnerable communities in developing countries. Fortunately, there is much that can be done now to reduce that vulnerability.

MANAGING RESIDUAL CLIMATE RISK

In the drylands of SSA and parts of Peninsular India, where much of the remaining hunger and poverty are concentrated, "the key challenge is to reduce water-related risks posed by high rainfall variability rather than coping with an absolute lack of water".

Yet the most viable opportunities for improving agricultural water management offer only incomplete control. A holistic strategy for investing in pro-poor agricultural water management requires parallel investment in other climate risk management measures to deal with the residual risk that water control alone cannot mitigate.

Climate risk management for agriculture includes:

- Systematic use of climate information and climate knowledge in strategic planning (*e.g.*, water control design, breeding) and adaptive decision making;
- Climate-informed technologies (*e.g.*, agricultural water management, drought-tolerant germplasm) and management strategies (*e.g.*, livelihood diversification) that reduce vulnerability to climate variability;
- Climate-informed policy (*e.g.*, early warning and response systems, safety nets) and market-based interventions (*e.g.*, insurance, credit) that transfer risk from vulnerable rural populations.

Climate risk management must address the full range of variability, balancing protection against the impacts of climatic extremes such as droughts and floods with effort to capitalize on opportunities arising from average and favourable climatic seasons. While improved water management is a crucial element, a portfolio of synergistic interventions is the most promising approach to covering the full spectrum of climate risks that confront farmers and impede the investment needed to realise the potential benefits of water management. Several options are available for managing the risk that feasible water management strategies cannot cover. A few – new ways to use new types of climate information, climateinformed livelihood strategies, innovations in financial risk transfer products – have not yet been fully explored or exploited.

MANAGING AGRICULTURE FOR RESILIENCE

In addition to investment in agricultural water management technologies, breeding for drought stress and diversification strategies can reduce vulnerability to moderate fluctuations in rainfall. Much of crop germplasm improvement targeting the tropical drylands is focused on resistance to drought and associated stresses. While drought-resistant germplasm development is fairly well established and supported relative to some of the emerging areas of climate risk management, there is still controversy about whether improving yields under drought must come at the expense of yields in seasons when rainfall is favourable.

Livelihood diversification can be an effective means to increase resilience in the face of climate variability if:

- The different income streams are not strongly correlated with each other or with seasonal rainfall,
- The diversified portfolio does not sacrifice substantial average income.

Opportunities for diversification can range from mixes of cultivars with staggered phenology at the field scale, to mixes of farm and non-farm enterprises across the household, to more diverse rural economies. Investment in rural roads, market infrastructure and education are likely to foster greater diversification.

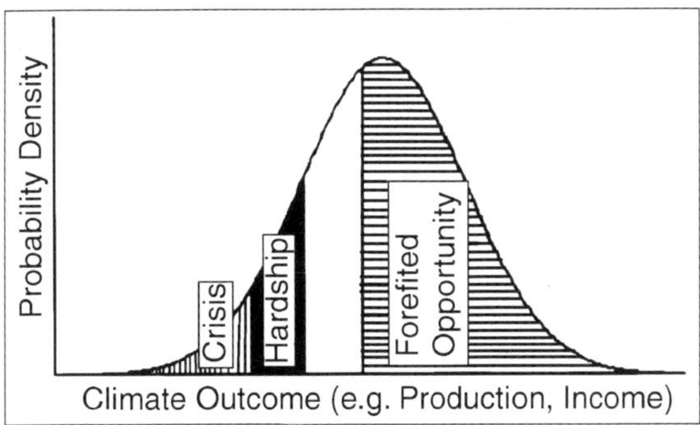

Fig. Idealized Representation of Impact of Climatic Risk Associated with Different Portions of the Distribution of Some Climate-sensitive Outcome.

Risk analysis, using historic climate data and standard analytical methods, can inform the design of more resilient diversified management strategies. There is also scope for such analyses to tailor germplasm development to small-scale water management such as conservation farming, and to mixes of cultivars that are less susceptible to dry spells than single cultivars.

USE OF CLIMATE INFORMATION BY RURAL COMMUNITIES

To the degree that climatic uncertainty adversely impacts farmer livelihoods, climate information that reduces uncertainty has potential to improve livelihoods. Relevant information includes historic climate records, monitoring of the current season, prediction at a range of lead times, and value-added information which integrates and translates raw climate data into impacts and management implications within agricultural and hydrological systems. Seasonal prediction is a particularly promising yet largely unexploited innovation in those regions and seasons where interactions between the atmosphere and underlying ocean and land surfaces provide a degree of predictability of rainfall at a seasonal lead time. The lead time of seasonal forecasts matches the period between the many climate-sensitive decisions (*e.g.*, allocation of farm land and household labour, choice of crops and production inputs, financing and procurement arrangements) that must be made prior to planting, and harvest when outcomes of those decisions are realised.

When the necessary conditions are in place, seasonal forecasts provide farmers the opportunity to adopt improved technology, intensify production, replenish soil nutrients and invest in more profitable enterprises when conditions are favourable or near average; and to more effectively protect their families and farms against the longterm consequences of adverse extremes.

Local historic climate records provide a means to quantify climate risk when making strategic investment and design decisions (*e.g.*, water storage or irrigation infrastructure), and when adapting new technologies to a local environment. Downscaling seasonal forecasts locally and interpreting inherently probabilistic forecasts in the context of historic variability require long-term records.

There is anecdotal evidence that biased perception of climate change – either underestimation of real change or overestimation of weak or non-existent trends – can contribute to poor management of rainfed agriculture. Routinely consulting climate observations provides a means to correct biases and to adjust management incrementally to adapt to progressive change. Real-time monitoring and prediction at a weather time scale contribute to food security and hydrometeorological (*e.g.*, flood) hazard early warning systems.

However information at these lead times are generally established within farmers' indigenous knowledge and observation, or accessible through existing services (*e.g.*, weather forecasts, agrometeorological bulletins). With few exceptions, the provision of climate information services to rural communities and other agricultural stakeholders (market institutions, government planners, food security response organizations) remains woefully inadequate throughout sub-Saharan Africa and India.

Seasonal climate forecasts have been disseminated routinely for the past ten years through regional climate outlook forums in much of Africa. Studies of the use and value of seasonal forecasts for smallholder agriculture reveals a high level of awareness and interest, and a range of promising management responses.

They have also identified obstacles associated with communication failures or resource constraints that limit capacity to respond. Constraints related to communication are, to a large degree, symptomatic of inadequate policies and institutional process, and therefore potentially amenable to intervention. Adoption rates and reported benefits have been fairly high in pilot projects in Zimbabwe, Burkina Faso and India and where extended interaction between smallholder farmers and researchers overcame some of the communication barriers.

The gap between current climate services in Africa and the needs of development is not limited to seasonal forecasts, although they have been the most studies. A multistakeholder, cross-sectoral assessment of the use of climate information in Africa concluded that (with a few noteworthy exceptions) the gap pervades across sectors and from local to policy levels, and attributed it in part to "market atrophy" resulting from the interplay between ineffective demand by development stakeholders and inadequate supply of relevant climate information services. Several gaps need to be addressed in parallel if climate information is to contribute to development at the scale of the MDGs.

INNOVATIONS IN FINANCIAL RISK TRANSFER

Recent innovations have resulted in a resurgence of effort to manage risk through insurance and other financial risk transfer instruments for smallholder rainfed agriculture in developing countries.

Basing insurance payouts on an objectively-measured index (*e.g.*, rainfall amount, modeled water stress, area-averaged production) that is correlated to loss instead of actual losses, overcomes problems with moral hazard (*i.e.*, incentive for farmers to let crops fail), adverse selection (less skilled farmers preferentially purchase insurance) and high transaction costs that have generally made traditional insurance unviable for smallholder farmers in developing countries.

However, it introduces basis risk: the residual uninsured risk that results from the imperfect relationship between the index on which payouts are based and the losses that the insurance is meant to protect against. Because index insurance does not protect against all risks, it must be carefully designed as a component of a comprehensive risk management policy.

Index insurance and associated financial products are being applied in several innovative, non-traditional ways to reduce particular climate risk-related constraints to agricultural development and rural poverty reduction. The World Bank's Commodity Risk Management Group is working with partners in several African and Latin American countries to implement bundled index insurancecredit- production input packages that overcome barriers to adoption of more profitable, intensified production technology by targeting the risk aversion of lenders.

The World Food Programme is working with the Ethiopian government on an index-based insurance project that would provide more timely access to funds to respond to emerging drought-related food crises. Barrett et al. provide a critical review of applications of index-based risk transfer products to reduce poverty, and propose a typology of applications based on where populations fall within a dynamic poverty trap framework.

Because some of the key obstacles to up-scaling index insurance for farmers and food crisis response relate to climate data, index-based insurance is increasing demand for climate information in data-sparse regions.

ANTICIPATING AND RESPONDING TO CLIMATERELATED CRISES

Climate variability imposes episodic extremes that even large-scale irrigation systems cannot mitigate, and that overwhelm the coping capacity of local populations without either outside intervention or high long-term cost to livelihoods. Advance estimates of staple crop production are an essential input to a range of contingency planning, food crisis management and market applications.

Early information about production shortfalls is a necessary but not sufficient condition for effective response, as delay in initiating response greatly increases the humanitarian and livelihood impacts of a food crisis, and the cost of aid. Because climate-related fluctuations in production can have large impacts on price and hence accessibility of staple foods, the use of production forecasts to manage markets and strategic grain reserves to stabilize prices is an appealing food security intervention.

The best-developed applications of climate information for risk management in SSA are for food security early warning and response. FEWSNET, JRC, FAO, WFP, AGRHYMET, SADC/RRSU and some countries have developed or routinely apply similar suites of spatial tools – rainfall monitoring, satellite vegetation monitoring, simple water balance-stress index models that incorporate historic and monitored weather data – to estimate food crop production shortfalls in advance of harvest.

These approaches have not yet successfully integrated the use of seasonal climate forecasting. For much of SSA especially, where there is potential forecasting skill in many regions and seasons, this represents a major opportunity to improve the lead time and consequently the performance of food security early warning systems.

The apparent increase in the frequency and extent of major flood events, whether due to land use changes, increasing vulnerability or climate changes, has created new interest in measures to reduce flood impacts. Beginning in 2003, the World Meteorological Organization convened a series of workshops to assess the status of flood forecasting systems throughout the world. Their findings indicated a wide range of capabilities.

The countries of west and central Africa had the most limited systems, due to shortages in real time data and observation networks, poor integration of climate data into hydrologic models and insufficient skills among weather and hydrologic services' staff. Some regional efforts are underway, through for example, AGRHYMET and the Niger Basin Authority, but since warning and responsc services are typically organized nationally, they do not replace the need for national level services.

In southern Africa the situation varies widely between countries, some with no operational systems and some pockets of good collaboration between meteorological and hydrological services. There is an absence of hydrologic forecasting in some countries for many of the same reasons that affect west and central Africa, including lack of data networks and skilled staff.

The report by WMO summarizes the key challenges for improving flood forecasting systems in all regions. Among the most prominent are the need for better observation systems and communication networks and the improvement of meteorological and hydrological forecasting products. Throughout Africa, there is a lack of flood risk mapping. Due to the dearth of hydrologic observation stations and damage to the few stations during flood events, there is no systematic record of flood events.

As a result, there is a lack of ability to estimate flood risk exposure or to map flood risk. Currently, only anecdotal evidence is used to estimate flood risk. Experience from the devastating floods of Mozambique provide further guidance on the importance of communication systems for transmission of flood forecasts, where operational flood forecasts actually exist. Mozambique has a history of extreme floods as it is positioned on the coast in the path of tropical storms and downstream of nine major river basins. There have been seven major floods since 1980 which instigated government and donor response for the development of operational flood forecasting. In 2000, seasonal forecasts shaded towards above average rainfall and flood warnings were issued. However, there were several sources of flood forecast information and in some cases they were contradictory. In general, communication to the public was poor.

Since then, investments have been made in hydrologic observation stations, community-based risk management and the mass media has been recognized as a key transmission route for flood warnings.

OPPORTUNITIES FOR NEW INVESTMENT

This report has reviewed the effect of climate variability on agricultural development in general and agricultural water management in particular. The findings indicate the limitations of singular strategies to manage the climate challenges to agriculture. While local water management can improve agricultural production in near normal years, it leaves farmers vulnerable to droughts and offers little advantage to exploit the opportunities of above normal rainfall years.

This may be a factor in why adoption rates are surprisingly low. Better results in agricultural development might be achieved by considering and managing the full spectrum of climate risk. A key underexploited opportunity is the use of climate information, including historical data, monitoring and prediction, to improve the design, operation and response of agriculture systems for resilience to climate variability and change.

In this section, we highlight three specific opportunities that we consider promising. Each targets a different layer of risk:

- Climate-informed investment in water management to increase the resilience of agricultural development and stimulate investment;
- Rural climate information services to support adaptive management of water and production activities, as a way to manage residual risk with incomplete water control;
- Integrated early warning systems to support more timely and better coordinated response to climatic shocks that exceed the coping capacity of rural communities.

ACCOUNTING FOR CLIMATEVARIABILITY IN SMALL-SCALE WATER MANAGEMENT

The case is clear for the need for better management of agricultural water resources in SSA and India, and in SSA especially, increasing the reliability of water supplies through water storage. Several studies have evaluated the cost effectiveness of small scale versus large scale storage investments, the reasons for low adoption of small scale storage methods, and the effects of irrigation on poverty.

However, there is little consideration of the optimal combination of large and small scale storage to meet the particular climate variability challenge for a river basin, region or nation from a macroscale (one example is a study of vulnerability that groups countries very broadly by rainfall season length and internal resources.

In fact, the use of climate information in these studies is lacking, as is any recognition that the climate challenges are significantly different depending on location and water demand.

Yet, these measures are all, in theory, a response to climate variability. Given the interest among donors and governments to increase investment in agriculture, and in water storage in particular, there is a major opportunity to coordinate and shape these investments such that they represent the best use of limited resources to manage the climate variability faced. A systems analysis approach to the design of water infrastructure was the hallmark of the Harvard Water Programme, which began as a multidisciplinary attempt to address irrigation problems in the Indus River valley.

The systems analysis approach, while still relevant, requires updating in its understanding of climate. Like all water management practices, the approach is based on the assumption that climate is stationary.

Additional innovations that can augment traditional approaches include the complementarities of small scale and large scale storage, the use of climate forecasts and monitoring, and the ability to manage risks financially through risk transfers at the microscale and macroscale.

An updated systems analysis approach to designing resilient water systems can be accomplished by identifying how climate affects agricultural production, characterizing the nature of climate dynamics at the scale of analysis, and using this information to create an integrated water management system that address the full spectrum of climate risks to agricultural development.

The process consists of the following steps:

- Assess current and historical impacts of climate variability on agricultural production and farmer welfare. This would be conducted using historical climate and economic data, which would be integrated in a GIS platform allowing the identification of climate risk "hotspots" and regions of enhanced vulnerability as well as opportunities.
- Characterize the hydroclimatic risk profile for the area of interest through analysis of historical climate data and model output. Topics to be addressed include: what are the probabilities and return periods of hazard causing climate events? Are there trends or periodicity? Is there codependency between spatial extent and severity of climate anomalies? Is there predictability on seasonal timescales or clear direction in climate change projections?
- Based on the understanding developed in the preceding steps, design an integrated water management system, grounded in the principles of integrated water resources management and developed in collaboration with stakeholders, for optimal investment. Water "supply curves," which represent an ordering of water supply options according to the unit cost for a given reliability will be calculated and

serve as guiding frame for investment selection within a systems analysis approach.

- Identify and manage residual risk through complementary use of non water interventions. Systematic analysis will identify remaining risks not covered by water management strategies. Non- water interventions for improving agricultural resilience are identified to address vulnerabilities. Examples include the use of climate information, including seasonal forecasts for identifying climate opportunities, and index insurance and early warning systems to protect against climate extremes.

The scarce consideration of climate is not unique to agricultural water management. In fact, the entire water management community is grappling with a response to growing understanding that climate is not stationary. A major part of that response should be based in the use of available climate information both in the planning and design of water management systems, as outlined above.

Currently a gap exists between operational water management and climate information. It will likely persist unless the use of climate information is integrated into water system design. This is vitally important in agricultural water management where infrastructure is lacking and climate is especially variable, *i.e.*, SSA, because the vulnerabilities are so widespread.

RURAL CLIMATE INFORMATION SERVICES

There is substantial opportunity for targeted investment to overcome known obstacles to delivering relevant climate information products and effective information and advisory services for rural communities where complete water control is not feasible in the foreseeable future. We recommend targeting investment to fill gaps across the full suite of interventions needed to make rural climate information systems work.

For countries in SSA, this would generally include parallel investment:

- Investment in climate data sets – potentially including digitizing paper archives, improving data storage and management, using remote-sensing to filling gaps in space and time, and negotiating data-sharing arrangements;
- Work with the NMS to design downscaled, tailored seasonal forecast products, and develop software tools and institutional capacity for sustained implementation.
- Development of curriculum and an in-service training programme in climate information communication and risk management for agricultural extension and effective NGO counterparts;
- Support for initial implementation of forecast training and dissemination workshops, evaluation of use and benefit, and participatory refinement of products and services at a pilot scale.

Depending on the country, some negotiation or investment in ICT may be needed to ensure a streamlined process for distributing climate information *e.g.*, to district agricultural administration offices.

A key outcome is agricultural extension services empowered to routinely convene the rural community (or their representatives) prior to the start of the growing season to present seasonal forecasts, discuss variability and any evidence of trends from the historic record, review previous season management and climatic outcomes for any lessons, discuss conditions monitored so far (*e.g.*, early rains, soil moisture, traditional indicators), and discuss management for the coming season.

We suggest investing initially at an intermediary scale that is large enough to foster significant change in institutions (NMS, NARES), and building careful impact evaluation into the effort from the onset. There is a tradeoff between investing at sufficient scale to have an impact on poverty and food security vs. testing interventions at a pilot scale long enough to collect credible evidence of benefit.

Degree of predictability (of the short rains) at a long lead time, and strength of regional institutions (*e.g.*, ICPAC, ASARECA) favour starting in the portion of eastern Africa that includes much of Kenya, Ethiopia, Uganda and northern Tanzania. The most promising opportunities are in semi-arid regions that are dominated by rainfed, cerealbased mixed farming systems. Kenya has a strong national agricultural research system (NARS).

The effectiveness of the formal agricultural extension system varies, and is often constrained by inadequate resources (*e.g.*, transportation, ICT infrastructure). The Arid Lands Resource Management Programme (ALRMP), supported by World Bank and housed in the Office of the President, has a mandate for drought management and capacity to serve an extension and communication role at at-least a pilot scale. The NMS hosts and collaborates closely with the regional climate centre (ICPAC).

Ethiopia also has a reasonably strong NARS. Recent decentralization and investment in rural development has strengthened the agricultural extension system. With support from Google.org, Ethiopia is preparing to develop long-term (30-year), high-resolution, merged satellite-station meteorological data sets intended to support development. Uganda has a similar fiscal and governance decentralization policy, and also benefits from strong farmer associations. We have less experience with Tanzania.

INTEGRATED EARLY WARNING SYSTEMS

Despite the best investments in agricultural water management and the best use of rural climate information services, a small number of extreme climate events will overwhelm the ability of farmers to cope. These events, though rare, can have devastating impacts on rural development efforts.

Early warning and response systems can help mitigate the impact of these events by providing farmers, government agencies and NGOs valuable time to prepare for and respond to extreme floods and droughts. However, the full potential of climate monitoring and forecasting has not yet been integrated with existing flood, drought and food security early warning systems.

Because the proportion of total uncertainty that is due to climate decreases through the growing season, the relative benefit of climate prediction (decreases in importance through the season) and monitoring (increases in importance) depends on the timing of the crop forecast. In most cases, accuracy (at a given lead time) or lead time (at a given level of accuracy) can be improved by incorporating additional information. Early warning systems often treat satellite vegetation indices (NDVI and its derivatives) as a complement to model-based estimates of yield loss from water stress.

Assimilating vegetation indices within model-based crop monitoring and forecasting systems can reconcile and optimally combine the two sources of information. Several methods are available for incorporating seasonal climate forecasts into model-based crop yield prediction.

Improvements in long-term, high-resolution, merged satellite-station data sets are expected to improve accuracy of model-based yield estimates. Increasing the length of rainfall proxy records that go into calibration will allow crop production monitoring and forecasts to be expressed in probabilistic terms, and the tradeoff between lead time and accuracy to be made explicit. In contexts where drought stress is not the dominant climatic cause of yield losses, modeling additional yield-limiting factors is expected to improve prediction.

The variety of crop production early warning systems that are operational complicates the decision about where to invest. In SSA, we prioritize regional technical institutions that have the political mandate to serve their member states.

In West Africa, the AGRHYMET Regional Centre is looking to make several enhancements to their operational crop monitoring system, and to expand their services from the CILSS (Sahelian) countries into all the countries that are part of ECOWAS. SADC/RRSU has the mandate in Southern Africa, but would need to be strengthened considerably as an institution before technical enhancements would be effective.

A WMO review of flood forecasting systems globally identified priority action areas for improving early warning systems. Prominent among these is the need for strengthening of observing and information systems. This includes meteorologic and hydrologic station networks, real-time data transmission and hydrometeorologic communication systems.

Satellite-based observation offers potential for improving identification of inundated areas and can complement hydrologic modeling, but forecasting depends on ground based observations for calibration and validation of predictions. The merging of ground based and satellite based observation systems is a promising opportunity for flood forecast improvement.

Further still, the combination of seasonal climate forecasts with lead times on the order of months with continuous monitoring for rapid detection of high risk areas has yet to be accomplished, although the technology exists. The lack of historical observations contributes to the uncertainty regarding flood risks throughout Africa.

There is no historical database of flood events, other than anecdotal, and as a result, no hydrologically-based maps of flood risk. Methods now exist to use historical observations of climate data, including global rainfall datasets and satellite-based digital elevation models in combination with hydrologic models to create synthetic historical estimates of flood risk and corresponding flood risk maps.

Such maps would be valuable in planning water storage investments and planning for warning systems and response. The lessons learned in Mozambique and the gaps in flood warning capabilities in many countries as outlined by the WMO reveal ample opportunity to reduce the risk of devastating floods on agricultural communities.

Characterization and mapping of flood risks, improved observation systems, development of flood forecasting skills and hazard communication systems are promising areas of investment for many poor countries in Africa and in regions of India, notably Bihar.

Without such efforts, farmers and many others remain vulnerable to floods striking without notice and destroying the hard won development gains accomplished through years of incremental improvements, such as agricultural water management.

There are a variety of opportunities to improve integrated early warning and response systems for hydroclimatological hazards. There are promising opportunities to exploit technological advances including remotely observed and modeled climate and environmental data. Priorities for investment include:

- Characterization of historical risk and uncertainties associated with hydroclimatological hazards.
- Merging of environmental monitoring and seasonal climate forecasting with modeling to improved forecasts of floods and famine.

How does Climate Variability Affect Agricultural Development

Climate variability directly affects crop production, primarily by driving supply of soil moisture in rainfed agriculture, and surface water run-off and shallow groundwater recharge in irrigated agriculture. Because biological response is non-linear and generally concave over some range of environmental variability, climate variability tends to reduce average yields.

Climate-driven fluctuations in production contribute substantially to volatility of food prices, particularly where remoteness, the nature of the commodity, transportation infrastructure, stage of market development or policy limit integration with global markets. Because market forces tend to move prices in the opposite direction to production fluctuations, variability in food crop prices tends to buffer farm incomes, but exacerbates food insecurity for poor net consumers.

The uncertainty associated with climate variability creates a moving target for management that reduces efficiency of input use and hence profitability, as management that is optimal for average climatic conditions can be far from optimal for growing season weather in most years. Crop responsiveness to fertilizer and planting density and hence optimal rates and profitability of production inputs, varies considerably from year to year as a function of water supply.

Climate variability and risk aversion on the part of decision makers cause substantial loss of opportunity in climatically-favourable seasons as a result of the precautionary strategies that vulnerable farmers employ *ex ante* to protect against the possibility of catastrophic loss in the event of a climatic shock.

Farmers' precautionary strategies – selection of less risky but less profitable crops, under-use of fertilizers, shifting household labour to less profitable off-farm activities, and avoiding investment in production assets, and improved technology – come at a substantial cost when climatic conditions are favourable.

Many of the coping responses that vulnerable households employ ex-post to survive an uninsured climate shock can have adverse, long-term livelihood consequences.

Coping strategies that include liquidating productive assets, defaulting on loans, migration, withdrawing children from school to work on farm or tend livestock, severely reducing nutrient intake and over-exploiting natural resources, even permanent abandonment of farms and migration to urban centers or refugee camps, sacrifice capacity to build a better life in the future.

What does Unmitigated Climate Risk Cost Agriculture

Estimating the impacts of climate variability and of specific climatic extremes is simpler at an aggregate (*e.g.*, national) scale than at household or community scales. Disentangling the various mechanisms of impact is even more difficult. Grey and Sadoff estimated that the occurrence of droughts and floods reduces Ethiopia's economic growth by more than one third.

Kenya suffered annual damages of 10-16 per cent of GDP due to flooding associated with El Niño in 1997-1998 and La Niña drought in 1998-2000. These damages extended beyond agriculture, with 88 per cent of flood losses incurred by the transport sector, while hydropower losses and industrial production totalled 84 per cent of the drought losses. There is substantial information about how smallholder farmers respond to risk, but few comprehensive empirical analyses of the livelihood impact of those responses.

For households within ICRISAT's village studies in India, Rosenzweig and Binswanger estimated that climate variability (expressed as timing of monsoon onset) accounts for a 15 per cent reduction of mean farm income for farmers within the median wealth class, and reduces income by 35 per cent for farmers in the lower quartile of wealth, indicating that less wealthy farmers were more willing than their relatively wealthy neighbours to sacrifice income to buffer themselves against climate variability.

A study of six villages in the Sahelian region of Burkina Faso showed that that variability of farm income (CV = 0.25) leads to food shortfalls in one out of every five years for the average farmer, but four out of five years for farmers in the bottom quartile of land holdings, and only once in ten years for the top quartile of farmers.

Relatively poor farmers in these villages forego about 18 per cent of their income to buffer against the existing level of risk (attributed primarily to climate variability), primarily by maintaining precautionary stores of grain, while the relatively wealthy farmers in the sample forego only 0.4 per cent of income.

Differentiating between the direct, ex-post impacts of climate-related shocks, and the cost associated with ex-ante decision-making in the face of climatic uncertainty is useful, as improved use of information has potential to reduce the latter.

Model-based studies that compare returns to profit-maximizing management with and without knowledge of weather for the upcoming season provide an estimate of the cost of uncertainty in the absence of risk aversion.

Climatic uncertainty costs the profit-maximizing farmer in Pergamino, Argentina, an estimated 23 per cent of gross margin on average, and reduces the efficiency of N fertilizer use by 39 per cent. For a semi-arid location in Kenya, we estimated that the inability to anticipate daily rainfall for the growing season costs the profit-maximizing farmer an average of 41 per cent of gross margin and 25 per cent of yield.

Elbers et al. present a unique attempt to quantify both the ex-post costs of climate fluctuations and the ex-ante opportunity cost due to climatic uncertainty and risk aversion to a set of farmers.

Combining 1980-2000 survey data from communal farm households in Zimbabwe with a dynamic household growth model, they attributed roughly one third of a 46 per cent risk-driven reduction in 50-year wealth accumulation to actual ex-post losses associated with climate and other fluctuations, while the remaining two-thirds was due to ex ante responses to the associated uncertainties.

ESTIMATING THE RELIABILITY OF RAINWATER HARVESTING

Rainwater harvesting (RWH) has a long lineage and has likely been practiced where ever rainfall was variable and needed. It is promoted as a climate change adaptation strategy and as a key strategy for better water management in developing countries.

However, the potential for RWH as an effective adaptation strategy has not been assessed except at the micro scale. Several analyses have considered the role of RWH to provide drinking water in urban areas including storage requirements and reliability estimation for particular locations. Brown et al. use global data sets to estimate the reliability of rainwater harvesting based on per capita water demand and gridded population data, at a resolution that is useful for assessing rainwater harvesting potential.

While the analysis cannot predict where RWH will succeed, it demonstrates the importance of the different timescales of rainfall variability for local water management interventions such as rainwater harvesting. Results reveals a wide area where rainwater harvesting has potential to provide the full storage needs of rainfed agriculture in Africa.

In these areas, total rainfall is sufficient to provide crop requirements, but intra-seasonal dry spells – which RWH is a well-suited to mitigate – may impact crops. In parts of Africa, notably in Ethiopia, Kenya and West Africa, where monthly rainfall in the average year is not sufficient to meet water requirements, RWH will still reduce susceptibility to intra-seasonal dry spells, but may not be sufficient to provide all the water needed. Other methods of mitigating rainfall variability, including large scale storage, may be more promising there.

PRIORITIES FOR STRENGTHENING CLIMATE INFORMATION SERVICES IN SSA

Institutional development to:

- Realign NMS from gatekeepers of data, to providers of services for development and participants in the development process;
- Ensure that products and services are driven by the needs of agriculture;
- Reform policy to make climate data available as a public good and resource for development;
- Revitalize and engage agricultural extension as providers of climate information and knowledge;
- Coordinate response between farmers and their advisers, market institutions, water managers and food crisis response organizations.

Observations and infrastructure investment to:

- Reverse decline in observing systems;
- Rescue and digitize paper records;

- Supplement sparse observations with merged remote sensing – station data sets;
- Enhance remote sensing products by extending duration (~30 years now possible), improving calibration, and correcting daily interpolation bias.

Information products designed (content, scale, format, timing) to meet known farmer needs; refined for particular contexts through ongoing stakeholder participation; with continuing investment in value-added products (*e.g.*, predictions of crop and forage yields, streamflow, flood hazard, pest and disease risk) for institutional users. Delivery mechanisms that integrate climate information as a routine part of agricultural extension (where functional); underpinned by training for intermediaries and by agricultural research; which foster farmer interaction and co-learning. Investment in rural communication infrastructure (radio, ICT) is also needed both as an alternative vehicle to reach rural communities and to streamline information transfer to communication intermediaries (*e.g.*, district agricultural offices). Evidence of development impact for rural climate information services, needed to mobilize resources and institutional support.

Weak body of evidence relative to other agricultural development interventions is due to:

- Lag time required to quantify impact particularly in the face of a stochastic driver,
- Widespread institutional failures that have constrained use beyond pilot projects,
- Neglect of impact assessment by proponents,
- Early studies that identified, but didn't seek to overcome, constraints to using existing operational products.

AGRICULTURAL IMPACTS ON SURFACE WATER

United Nations' predictions of global population increase to the year 2025 require an expansion of food production of about 40-45 per cent. Irrigation agriculture, which currently comprises 17 per cent of all agricultural land yet produces 36 per cent of the world's food, will be an essential component of any strategy to increase the global food supply. Currently 75 per cent of irrigated land is located in developing countries; by the year 2000 it is estimated that 90 per cent will be in developing countries.

In addition to problems of waterlogging, desertification, salinization, erosion, etc., that affect irrigated areas, the problem of downstream degradation of water quality by salts, agrochemicals and toxic leachates is a serious environmental problem. "It is of relatively recent recognition that salinization of water resources is a major and widespread phenomenon of possibly even greater concern to the sustainability of irrigation than is that of the salinization of soils, per se. Indeed, only in the past few years has it become apparent that trace toxic constituents, such as Se, Mo and As in agricultural drainage waters may cause pollution problems that threaten the continuation of irrigation in some projects".

PUBLIC HEALTH IMPACTS ON POLLUTED WATER

Polluted water is a major cause of human disease, misery and death. According to the World Health Organization (WHO), as many as 4 million children die every year as a result of diarrhoea caused by water-borne infection. The bacteria most commonly found in polluted water are coliforms excreted by humans. Surface run-off and consequently non-point source pollution contributes significantly to high level of pathogens in surface water bodies. Improperly designed rural sanitary facilities also contribute to contamination of groundwater.

Agricultural pollution is both a direct and indirect cause of human health impacts. The WHO reports that nitrogen levels in groundwater have grown in many parts of the world as a result of "intensification of farming practice". This phenomenon is well known in parts of Europe.

Nitrate levels have grown in some countries to the point where more than 10 per cent of the population is exposed to nitrate levels in drinking water that are above the 10 mg/l guideline. Although WHO finds no significant links between nitrate and nitrite and human cancers, the drinking water guideline is established to prevent methaemoglobinaemia to which infants are particularly susceptible.

Although the problem is less well documented, nitrogen pollution of groundwater appears also to be a problem in developing countries.

Lawrence and Kumppnarachi reported nitrate concentrations approaching 40-45 mg N/l in irrigation wells that are located close to the intensively cultivated irrigated paddy fields. The variation in NO_3-N which shows a peak in the *maha* (main) cropping season when rice growing is most intensive in Sri Lanka.

Reiff, irrigated agriculture, notes that water pollution is both a cause and an effect in linkages between agriculture and human health.

The following health impacts which apply, in particular, to developing countries, were noted by Reiff:

- Adverse environmental modifications result in improved breeding ground for vectors of disease (*e.g.* mosquitos). There is a linkage between increase in malaria in several Latin American countries and reservoir construction. Schistosomiasis (Bilharziasis), a parasitic disease affecting more than 200 million people in 70 tropical and subtropical countries, has been demonstrated to have increased dramatically in the population following reservoir construction for irrigation and hydroelectric power production. Reiff indicates that the two groups at greatest risk of infection are farm workers dedicated to the production of rice, sugar cane and vegetables and children that bathe in infested water.

- Contamination of water supplies primarily by pesticides and fertilizers. Excessive levels of many pesticides have known health effects.
- Microbiological contamination of food crops stemming from use of water polluted by human wastes and run-off from grazing areas and stockyards. This applies both to use of polluted water for irrigation and by direct contamination of foods by washing vegetables etc. in polluted water prior to sale. In many developing countries there is little or no treatment of municipal sewage, yet urban wastewater is increasingly being used directly or recycled from receiving waters, into irrigated agriculture. The most common diseases associated with contaminated irrigation waters are cholera, typhoid, ascariasis, amoebiasis, giardiasis and enteroinvasive *E. coli*. Crops that are most implicated with spread of these diseases are ground crops that are eaten raw such as cabbage, lettuce, strawberries, etc.
- Contamination of food crops with toxic chemicals.
- Miscellaneous related health effects, including treatment of seed by organic mercury compounds, turbidity (which inhibits the effectiveness of disinfection of water for potable use), etc.

To this list can be added factors such as the potential for hormonal disruption (endocrine disruptors) in fish, animals and humans. Hormones are produced by the body's endocrine system.

Because of the critical role of hormones during early development, toxicological effects on the endocrine system often have impacts on the reproductive system.

While pesticides such as DDT have been implicated, the field of endocrine disruption is in its infancy and data which support cause and effect are not yet conclusive.

It is probably safe to conclude, however, that high levels of agricultural contaminants in food and water as are found in many developing country situations have serious implications for reproduction and human health.

AGRICULTURE AND THE ARAL SEA DISASTER

The social, economic and ecological disaster that has occurred in the Aral Sea and its drainage basin since the 1960s, is the world's largest example of how poorly planned and poorly executed agricultural practices have devastated a once productive region.

AGRICULTURAL WATER POLLUTION IN DEVELOPING COUNTRIES

Data on water pollution in developing countries are limited. Further, such data are mostly "aggregated", not distinguishing the relative proportion of "point" and "non-point" sources.

In Thailand, the Ministry of Public Health reported the results of pollution monitoring of 32 rivers. Pesticide consumption has strongly increased in all developing countries. In India, consumption increased nearly 50-fold between 1958 and 1975. Yet the Indian consumption in 1973-74 was reported to be averaging a mere 330 g/ha, compared to 1483 g in USA and 1870 g in Europe. According to various surveys in India and Africa, 20-50 per cent of wells contain nitrate levels greater than 50 mg/l and in some cases as high as several hundred milligrams per litre. In the developing countries, it is usually wells in villages or close to towns that contain the highest levels, suggesting that domestic excreta are the main source, though livestock wastes are particularly important in semi-arid areas where drinking troughs are close to wells.

DECISIONS IN AGRICULTUREFOR NON-POINT SOURCE POLLUTION CONTROL

Decisions by agriculturalists for control of agricultural non-point source pollution can be at various scales. At the field level, decisions are influenced by very local factors such as crop type and land use management techniques, including use of fertilizers and pesticides.

These decisions are based on best management practices that are possible under the local circumstances and are meant to maximize economic return to the farmer while safeguarding the environment.

Local decisions are made on the basis of known relationships between farm practice and environmental degradation but do NOT usually involve specific assessment of farm practices within the larger context of river basin impacts from other types of sources.

Decisions regarding use of waste water, sludges, etc., for agricultural application are also made using general knowledge of known impacts and of measures to mitigate or minimize these impacts. However, the challenge for agriculturists is to mobilize the knowledge base and to make it available to farmers. At the river basin scale, the nature of decision making is quite different. At this scale, the typical decision-making problem for non-point source control in many developing countries.

It is not possible in this publication to describe in detail the "tools" that are used to address this basin-scale management problem. Moreover, many of the tools are not yet systematized to the point where they are easily accessible to agricultural practitioners.

THE DATA PROBLEM

One area, however, that is well known, is the data problem. The water quality database that is available in many developing countries (and in some developed countries) is of little value in pollution management at the river basin scale nor is it useful for determining the impact of agriculture relative to other types of anthropogenic impacts.

A common observation amongst water quality professionals is that many water quality programmes, especially in the developing countries, collect the wrong parameters, from the wrong places, using the wrong substrates and at inappropriate sampling frequencies and produce data that are often quite unreliable. Further, the data are not assessed or evaluated and are not sufficiently connected to realistic and meaningful programme, legal or management objectives. This is not the fault of the developing countries; more often it results from inappropriate technology transfer from the developed countries and an incorrect assumption by recipients and donors that the data paradigm developed by the developed countries is appropriate in the developing countries.

Additionally, water quality monitoring programmes, worldwide, are under severe stress as governments reduce budgets, downsize and shift priorities. "Monitoring" has become a dirty word and governments are increasingly reluctant to pay for it. Paradoxically, the need for reliable water quality information has never been greater. Fortunately, new scientific research, together with budget realities, now makes it possible to rethink and redesign data programmes that are inherently more focused, more practical, more efficient, produce more information and less data and which meet programme goals in measurable economic terms.

This publication is not the place to deal substantively with new monitoring (data collection) techniques; however, it is sufficient to note here that monitoring technology has changed dramatically in the past decade, to the point where significant economic and information gains can be achieved in most monitoring programmes. Significant for agricultural programmes is that water quality data are rarely collected by ministries of agriculture. Nevertheless, sustainable agriculture within the framework of comprehensive basin management will require relevant and reliable data upon which to make management decisions. This will necessitate intervention by agriculturalists in existing water quality data programmes if relevant data are to be collected for agricultural management purposes.

POLLUTION BY SEDIMENTS

Although agriculture contributes to a wide range of water quality problems, anthropogenic erosion and sedimentation is a global issue that tends to be primarily associated with agriculture. While there are no global figures, it is probable that agriculture, in the broadest context, is responsible for much of the global sediment supply to rivers, lakes, estuaries and finally into the world's oceans.

Pollution by sediment has two major dimensions. One is the Physical Dimension - top soil loss and land degradation by gullying and sheet erosion and which leads both to excessive levels of turbidity in receiving waters and to off-site ecological and physical impacts from deposition in river and lake beds. The other is a Chemical Dimension - the silt and clay fraction (<63 m m fraction), is a primary carrier of adsorbed chemicals, especially phosphorus, chlorinated pesticides and most metals, which are transported by sediment into the aquatic system.

Erosion is also a net cost to agriculture insofar as loss of top soil represents an economic loss through loss of productive land by erosion of top soil and a loss of nutrients and organic matter that must be replaced by fertilizer at considerable cost to the farmer in order to maintain soil productivity. The reader is referred to Roose for a detailed analysis of the social, economic and physical consequences of erosion of agricultural land and of measures that should be taken to control erosion under different types of land use, especially in developing countries. Whereas Roose is mainly concerned with the impact of erosion on agriculture, this publication is primarily concerned with agricultural erosion from the perspective of its impacts on downstream water quality.

Control of agricultural pollution usually begins, therefore, with measures to control erosion and sediment run-off. Therefore,the principal mechanisms which govern erosion processes and those measures which can be taken to control erosion.

SEDIMENT AS A PHYSICAL POLLUTANT

Global estimates of erosion and sediment transport in major rivers of the world vary widely, reflecting the difficulty in obtaining reliable values for sediment concentration and discharge in many countries, the assumptions that are made by different researchers and the opposing effects of accelerated erosion due to human activities (deforestation, poor agricultural practices, road construction, etc.) relative to sediment storage by dam construction.

Milliman and Syvitski estimate global sediment load to oceans in the mid-20th century at 20 thousand million t/yr, of which about 30 per cent comes from rivers of southern Asia. Significantly, they believe that almost 50 per cent of the global total comes from erosion associated with high relief on islands of Oceania - a phenomenon which has been underestimated in previous estimates of global sediment production. While erosion on mountainous islands and in upland areas of continental rivers reflects natural topographic influences, Milliman and Syvitski suggest that human influences in Oceania and southern Asia cause disproportionately high sediment loads in these regions.

Sediment, as a physical pollutant, impacts receiving waters in the following principal ways:

- High levels of turbidity limit penetration of sunlight into the water column, thereby limiting or prohibiting growth of algae and rooted aquatic plants. In spawning rivers, gravel beds are blanketed with fine sediment which inhibits or prevents spawning of fish. In either case, the consequence is disruption of the aquatic ecosystem by destruction of habitat. Notwithstanding these undesirable effects, the hypertrophic (nutrient rich) status of many shallow lakes, especially in developing countries, would give rise to immense growth of algae and rooted plants were it not for the limiting effect of light extinction due to high turbidity. In this sense, high turbidity can be "beneficial" in highly eutrophic lakes; nevertheless, many countries recognise that this situation is undesirable for both aesthetic and economic reasons and are seeking means to reduce both turbidity and nutrient levels. Box 4 presents the impact of sediment on coral reefs.
- High levels of sedimentation in rivers leads to physical disruption of the hydraulic characteristics of the channel. This can have serious impacts on navigation through reduction in depth of the channel and can lead to increased flooding because of reductions in capacity of the river channel to efficiently route water through the drainage basin. For example, calculations by the UFRGS (1991) of erosion and sediment transport in the Sao Francisco River Basin, a large drainage system in eastern Brazil, demonstrate that the central portion of the river basin is now dominated by sediment deposition. This has resulted in serious disruption of river transportation and clogs hydraulic facilities that have been built to provide irrigation water from the main river channel. The sediment largely originates from rapidly eroding sub-basins due to poor agricultural practices.

Sediment as a Chemical Pollutant

The role of sediment in chemical pollution is tied both to the particle size of sediment and to the amount of particulate organic carbon associated with the sediment. The chemically active fraction of sediment is usually cited as that portion which is smaller than 63 m m (silt + clay) fraction. For phosphorus and metals, particle size is of primary importance due to the large surface area of very small particles. Phosphorus and metals tend to be highly attracted to ionic exchange sites that are associated with clay particles and with the iron and manganese coatings that commonly occur on these small particles. Many of the persistent, bioaccumulating and toxic organic contaminants, especially chlorinated compounds including many pesticides, are strongly associated with sediment and especially with the organic carbon that is transported as part of the sediment load in rivers. Measurement of phosphorus transport in North America and Europe indicate that as much as 90 per cent of the total phosphorus flux in rivers can be in association with suspended sediment.

The affinity for particulate matter by an organic chemical is described by its octanol-water partitioning coefficient (K_{OW}). This partitioning coefficient is well known for most organic chemicals and is the basis for predicting the environmental fate of organic chemicals. Chemicals with low values of K_{OW} are readily soluble, whereas those with high values of K_{OW} are described as "hydrophobic" and tend to be associated with particulates. Chlorinated compounds such as DDT and other chlorinated pesticides are very hydrophobic and are not, therefore, easily analysed in water samples due to the very low solubility of the chemical. For organic chemicals, the most important component of the sediment load appears to be the particulate organic carbon fraction which is transported as part of the sediment. Scientists have further refined the partitioning coefficient to describe the association with the organic carbon fraction (K_{OC}).

Another important variable is the concentration of sediment, especially the <63 m m fraction, in the water column. Even those chemicals that are highly hydrophobic will be found in trace levels in solublc form. Where the suspended load is very small (less than 25 mg/l), the amount of water is so large relative to the amount of sediment that the bulk of the load of the chemical may be in the soluble fraction. This becomes an important issue in the monitoring of hydrophobic chemicals.

Unlike phosphorus and metals, the transport and fate of sediment-associated organic chemicals is complicated by microbial degradation that occurs during sediment transport in rivers and in deposited sediment.

Nevertheless, the role of sediment in the transport and fate of agricultural chemicals, both for nutrients, metals and pesticides is well known and must be taken into account when monitoring for these chemicals and when applying models as a means of determining optimal management strategies at the field and watershed level.

For this reason, models using the "fugacity" concept (uses the partitioning characteristics of chemicals as a basis for determining the environmental compartment - air, sediment, water, biota - in which the chemical is primarily found) has proven effective in predicting the environmental pathways and fate of contaminants.

Organic chemicals associated with sediment enter into the food chain in a variety of ways.

Sediment is directly ingested by fish however, more commonly, fine sediment (especially the carbon fraction) is the food supply for benthic (bottom dwelling) organisms which, in turn, are the food source for high organisms. Ultimately, toxic compounds bioaccumulate in fish and other top predators. In this way, pesticides that are transported off the land as part of the run-off and erosion process, accumulate in top predators including man.

KEY PROCESSES: PRECIPITATION AND RUN-OFF

The major characteristic of non-point source pollution is that the primary transfer mechanisms from land to water are driven by those hydrological processes that lead to run-off of nutrients, sediment and pesticides. This is important, not only to understand the nature of agricultural pollution, but also because modelling of hydrological processes is the primary mechanism by which agriculturalists estimate and predict agricultural run-off and aquatic impacts. Except where agricultural chemicals are dumped directly into watercourses, almost all other non-point source control techniques in agriculture involve control or modification of run-off processes through various land and animal (manure) management techniques.

In large parts of the world, precipitation is in the form of rain. However, in those areas where precipitation is in the form of snow, the science becomes more complex. Nevertheless, control measures, whether for areas subject to rain or snow can be easily summarized. Therefore, for the purpose of this publication, focus will be on the relationships between rainfall and run-off.

While the practice of hydrology can be quite theoretical, the principal concepts are easily understood.

Rainfall

The primary controlling factor is the rate (intensity) of rainfall. This controls the amount of water available at the ground surface and is closely related to measures of energy that are used in many mathematical formulations to calculate soil detachment by rain drops. Soil detachment makes soil particles available for sediment run-off.

Soil Permeability

Permeability is a physical characteristic of a soil and is a measure of the ability of the soil to pass water, under saturated conditions, through the natural voids that exist in the soil. Permeability is a function of soil texture, mineral and organic composition, etc.. In contrast, "porosity" is the measure of the amount of void space in a soil; however, permeability refers to the extent to which the porosity is made up of interconnecting voids that allow water to pass through the soil. As an example, styrofoam is highly porous but impermeable, whereas a sponge is both porous and permeable.

Infiltration

Infiltration rate, the rate at which surface water passes into the soil (cm/hr), is one of the most common terms in hydrologic equations for calculating surface run-off. Infiltration is not identical to permeability; it is mainly controlled by capillary forces in the soil which, in turn, reflect the prevailing conditions of soil moisture, soil texture, degree of surface compaction, etc. Infiltration will vary between and within rainfall events, depending upon factors such as antecedent soil moisture, nature of vegetation, etc. In general, infiltration rate begins at a high value during a precipitation event and decreases to a small value when the soil has become saturated.

Surface Run-off

This is the amount of water available at the surface after all losses have been accounted for. Losses include evapotranspiration by plants, water that is stored in surface depressions caused by irregularity in the soil surface and water that infiltrates into the soil.

The interaction between infiltration rate and precipitation rate mainly governs the amount of surface run-off. Intense rainstorms tend to produce much surface run-off because the rate of precipitation greatly exceeds the infiltration rate.

Similarly, in areas of monsoonal rain and tropical storms, the length and intensity of precipitation frequently exceeds infiltration capacity. Destruction of protective surface vegetation and compaction of the soil, especially in tropical environments, leads to major erosional phenomena due to the amount of surface run-off. Except for nitrogen which is usually found in groundwater in agricultural areas, surface run-off is the primary contributor of agricultural chemicals, animal wastes and sediment to river channels.

Interflow

(sometimes called "throughflow") Because soil horizons have different levels of permeability not all water in the soil will move downward into the groundwater. The residual water in the soil will move along the soil horizons, parallel to the ground surface. Interflow usually emerges near the bottom of slopes and in valley bottoms.

Therefore, identification of these hydrologically active zones is an important part of agricultural non-point source control measures. Interflow is the mechanism which has also been linked to soil piping, a potentially destructive characteristic in some soils by which shallow "pipes" form naturally in the soil and are enlarged by interflow to the point where they collapse causing gullies in the agricultural surface.

Groundwater

Groundwater is supplied by water which passes through the soil horizons into the parent material and/or bedrock underlying the soil. Groundwater tends to flow towards rivers channels where it emerges and supports stream flow during periods of little or no rain. This component of stream flow is called "base flow". The chemistry of base flow reflects the soil and bedrock geochemistry, plus any agrochemicals that have been leached into the groundwater.

Snowmelt

The phenomenon of snowmelt greatly complicates prediction of agricultural pollution using conventional hydrologic models. Snowmelt, by itself, is not normally a major producer of surface run-off. However, the combination of spring rain and snowmelt on frozen or thawing soils can produce serious erosional problems. Snowmelt tends to contribute greatly to agricultural non-point source pollution by carrying to adjacent streams the animal wastes, sludges and other wastes that were spread on frozen agricultural soils during the winter period. Correct management of animal wastes in regions of frozen ground has major beneficial effects on water quality.

SEDIMENT DELIVERY RATIO

The sediment delivery ratio (SDR) is commonly used in erosion and transport studies to describe the extent to which eroded soil (sediment) is stored within the basin. The SDR is defined as:

$$\text{SDR} = \frac{\text{Measured Sediment Yield}}{\text{Grosserosionin the basin}}$$

where *yield* is determined from reservoir sedimentation or from a sediment monitoring station and*gross erosion* is estimated using an estimation techniques such as the Universal Soil Loss Equation.

The SDR is always less than 1.0, indicating that soil that is eroded at the field level tends not to travel far before it is deposited. Indeed, sediment storage in rills on fields, at field margins and at the foot of slopes is large. Storage also occurs in river channels (bed and overbank deposition), in wetlands and in reservoirs and lakes. The SDR is highly variable, however the concept is one of the most important in the understanding of erosion and sedimentation processes and how these operate in time and space.

SEDIMENT ENRICHMENT RATIO

The concept of the sediment enrichment ratio (SER) is quite important in understanding the impact and economic cost of chemical loss from fields. The process of surface erosion tends to be selective towards fine particles. Consequently, the particle size characteristics of material eroded at source (at the plot level) is progressively changed towards finer particles through deposition of the coarser fraction (*e.g.* sand-size material). Because of the chemically enriched nature of fine particles due to the large surface area of clay-size sediment, the concentration of chemicals that are associated with sediment (phosphorus, metals, organic nitrogen, hydrophobic pesticides) increases as the impoverished sand-size fraction is lost during down-field transport resulting in an increasing proportion of the chemically enriched fine (silt-clay) fraction.

The Sediment Enrichment Ratio (SER) is defined as:

$$\text{SER} = \frac{\text{Concentration of chemical "X" in transported sediment}}{\text{Concentration of the chemical "X" in the soil}}$$

Sediment chemistry is measured at some point downslope, *e.g.*, at the edge of a field or in adjacent streams.

The importance of the enrichment ratio lies in the fact that there is proportionally more fine-grained sediment transported than coarse-grained sediment during surface erosion. Therefore, the sediment being transported has a finer texture than the source soil material. Because of the affinity of soil nutrients for fine sediment, this proportionally larger loss of fine material means that there is net impoverishment of the soil. (Fertilizers), this usually constitutes "mining" of the natural nutrition of the soil (often referred to as "natural capital") and which may never be replaced by the addition of fertilizer. The cost to the farmer is therefore two-fold: loss of productivity due to loss of natural nutrition in the soil; and economic cost of fertilizer which is added in the attempt to compensate for this loss.

MEASUREMENT AND PREDICTION OF SEDIMENT LOSS

Prediction Models

Agriculturalists worldwide have spent much time and resources attempting to find reliable methods of predicting erosion and sediment-associated chemical run-off under different conditions of crop type, tillage practices, etc. Consequently, there is a large number of models that have been developed for the prediction of agricultural non-point source run-off of sediment, nutrients and pesticides. Many of the models permit gaming with alternative choices of land management, crop type and fertilizer and pesticide application rates. Because all models (except unit load models) require hydrometric input and many use a sediment sub-component, it is appropriate to integrate these into a single table together with their principal characteristics.

In general, there are three types of models based on input data needs:

- Simple screening models, such as the unit load approach, attempt to provide approximate answers about the likely magnitude of sediment and chemical run-off. This approach is a statistical methodology in which data on run-off of sediment, nutrients and pesticides on a unit area basis (*e.g.* tonnes of sediment per hectare) are collated from many studies to reflect similarities in crop type, soil and physiographic characteristics. Unlike the other types of models which are largely focused on prediction and improvement of agricultural management at the farm level, the unit load approach is mainly focused on impacts of agriculture on downstream water quality and without consideration of alternative farm management practices.

- Simple empirical relationships: the widely used and respected Universal Soil Loss Equation (USLE) of Wischmeier (1976) has had remarkable success at the plot level and has been incorporated into many of the complex models. The USLE is designed as a field management tool and provides aggregated information at the storm, seasonal or annual level. Wischmeier (1976) reported that the average prediction error for annual soil loss was 12 per cent; larger errors are to be expected for single storm events. The USLE is one way to determine erosion potential for input into the denominator of the Sediment Delivery Ratio. The USLE is detailed here because of its success and because the same type of approach has been used in Africa and elsewhere.The USLE is calculated as:

$$A = R.\ K.\ LS.\ C.\ P$$

where:

A = Calculated soil loss in t/ha for a given storm or period.
R = Rainfall energy factor.
K = Soil erodibility factor.
LS = Slope-length factor.
C = Cropping management factor (vegetative cover).
P = Erosion-control practice factor.

Each of the factors can be calculated or estimated using field data and from tables or nomograms for all other factors. Novotny and Olem provide an excellent commentary on this and other methods for estimating or modelling erosion. The USLE is designed for rainfall only and does not handle snowmelt or rainfall on frozen ground. The USLE requires calibration data from standard plot experiments which are widely available in North America and, more limitedly, from other parts of the world.

At the international level, the simplicity and effectiveness of the soil loss approach prompted an important series of experiments in Zimbabwe in the 1950s and 1960s with the primary objective of determining losses of nitrogen, phosphorus and organic carbon from the natural fertility of the soil. According to Stocking who exhaustively analysed this database, it represented the "best such database in any developing or tropical country". This work led to the development of the SLEMSA model for conditions in Southern Africa. The value of this approach has further led FAO to develop an international network, "Network on Erosion-Induced Loss in Soil Productivity", with research partners in Africa, South America and Asia.

Although Wischmeier has cautioned against extending soil loss models beyond field loss studies, these models are intuitively attractive for predicting erosion over large areas. It should be noted that, due to transport losses, such erosion estimates apply only to total erosion at source and do not reflect sediment loads (or sediment yield) measured at downstream locations. Such estimating techniques would, if calibrated so that the errors are known, have useful application as a screening tool for the estimation of erosion potential under conditions of similar crop, soil and topographic factors over large areas. Internationally, there appears to be little systematic information on calibration; however the Network on Erosion-Induced Loss in Soil Productivity may eventually provide suitable information.

The most extreme example of use of the soil loss approach for large area estimation is in Brazil where the UFRGS and Carvalho used large area maps and satellite data to estimate several of the parameters of the USLE for application at regional scales. The intent was to provide generalized estimates of regional erosion potential for the entire country.

While this approach has wide margins of error, it represents a method of screening for major change in erosion potential arising from combinations of agricultural land use, climate and topography and merits further consideration, especially where plot calibration and in-river sediment monitoring data are available.

Together with the need to further develop screening level models, is the need to generate improved field data on erosion and sediment loss. Hudson presented a wide range of simple field measurement techniques that are particularly useful in developing countries.

There is a wide variety of deterministic and stochastic models that attempt to simulate the physics of the erosion process. The data requirements for calibration and verification are extremely large. While such models may have certain advantages, especially in terms of the level of detail in which one can simulate alternative farm practices, these are generally unsuitable for developing countries due to their data requirements and the observation that management judgements for farm-level decisions can almost always be made on the basis of more generalized data combined with experience and common sense.

Sediment Yield

Sediment yield, usually expressed as tonnes per unit area of the basin per year, is the amount of sediment measured at some point in the basin divided by the basin area. It always less than the total erosion due to sediment storage during transport and is highly variable because of measurement difficulty, the temporal variability of hydrological processes and changes in land management practices in the basin from one year to the next.

The Sediment Delivery Ratio subsumes a wide range of storage processes that operate at the field, sub-basin and basin scales. Sediment yield values are also highly variable in time, with smaller basins subject to greater variability than larger basins. Consequently, published values for sediment yield (t/ha/yr) must be interpreted with great caution.

Estimates of sediment yield have important economic consequences. In many developing countries the database with which to estimate reservoir life is very limited. According to White examples of predicted sediment yield in Asia tend to be between two and sixteen times lower than actual measured rates, with the consequence that actual reservoir life is greatly reduced. This is also a problem in northern Africa. Partly this arises from the use of unreliable prediction techniques and the use of short-term data which usually fail to account for occasional but severe episodes of erosion (*e.g.* major storm events) and from increased land pressure after construction of the reservoir. White reports that the Magat Dam was constructed on the basis of a planned sediment yield of 20 t/ha/yr when, in reality, the yield was shown to be 38 t/ha/yr - cutting the useful life of the reservoir in half! In the Republic of South Africa, off-site damage due to soil erosion was estimated in 1989 to total US$ 37.6 million annually. Because agriculture is a major contributor to sediment yield it is essential that national agricultural organizations account for off-site costs.

Scale Problems

In all aspects of monitoring, modelling, or prediction of non-point source management, the scale factor is poorly understood by many practitioners. The scale factor not only pertains to the cost of collecting data for model calibration, but it is also vital to the ability to extrapolate useful management principles that will apply to larger areas. As an example, during the 1970s it was assumed by many modellers who were engaged in the non-point source management programme of the North American Great Lakes, that extrapolation of cause-effect relationships from small-area studies into larger parts of the Great Lakes basin was feasible. It was found, however, that extrapolation became quite uncertain as basin area increased, for a variety of reasons.

The problem of extrapolation has not been definitively addressed in regards to levels of uncertainty. For example, although it is known that control of agricultural erosion is essential to mitigate nutrient (especially phosphorus) loss to receiving watercourses, it is unclear over what distance phosphorus run-off from agriculture has an immediate impact.

The question then is, does erosion control in far upstream areas have an immediate benefit for a receiving water that may lie hundreds of kilometers downstream? Alternatively, one may adopt the position that over longer periods of time (scores of years) large storms will ensure that most of the eroded sediment is, in fact, transported through the river system and, therefore, one can assume a 1:1 relationship between upstream sediment production and downstream sediment transport. Research by Meade and Trimble and others shows that over longer time periods, sediment is mobilized differentially along the river basin. Contemporary sediment mobilization in the middle reaches of piedmont rivers of the United States represented sediment that was eroded several decades earlier but which had been deposited during transport into long-term storage as valley floor deposits.

The essential point is that time and space scales must be recognized when preparing management plans for erosion control. While the near-term physical benefits of erosion control are likely to be felt quickly in receiving waters, sediment-associated contaminants that are stored in the river basin may take decades to finally be transported out of the basin, notwithstanding the resources spent on upstream erosion control.

MANAGEMENT OF WATER QUALITY IMPACTS FROM FERTILIZERS

The investigation of eutrophication of surface water by agriculture must adopt a pragmatic management perspective. Of value to agriculture is the perspective adopted by the OECD study of eutrophication.

That study focused on the following aspects:

- The qualitative assessment of the trophic state of bodies of water in terms of a few easily measured parameters.
- The dependence of this state on nutritional conditions and nutrient load.
- Translation of these results to the needs of eutrophication control for management.

The progression of these aspects is interesting in that the focus is on easily measurable state of the water body, followed by a determination of the extent to which the state is a product of nutrient loads, then the degree to which loads may be manipulated to achieve a desired trophic state that is determined by water use. Prediction of water quality impacts of fertilizers and related land management practices is an essential element of site-specific control options and for the development of generic approaches for fertilizer control. Prediction tools are essentially in the form of models.

Mineral Fertilizers

The response to the need to control leaching and run-off of nutrients and contamination of soils and water by heavy metals has been variable in Europe. Control measures are part of the larger issue of mineral and organic fertilizer usage. FAO/ECE (1991) summarized the types of voluntary and mandated controls in Europe that apply to mineral fertilizers as:

- Taxes on fertilizer.
- Requirement for fertilizer plans.
- Preventing the leaching of nutrients after the growing season by increasing the area under autumn/winter green cover and by sowing crops with elevated nitrogen

- Promoting and subsidizing better application methods, developing new, environmentally sound fertilizers and promoting soil testing.
- Severely limiting the use of fertilizers in *e.g.* water extraction areas and nature protection areas.

In any location where intensive agriculture and/or livestock farming produces serious risks of nitrogen pollution, Ignazi (1993) recommended the following essential steps that are taken at the farm level:

Rational Nitrogen Application

To avoid over-fertilization, the rate of nitrogen fertilizer to be applied needs to be calculated on the basis of the "crop nitrogen balance". This takes into account plant needs and amount of N in the soil.

Vegetation Cover

As far as possible, keep the soil covered with vegetation. This inhibits build-up of soluble nitrogen by absorbing mineralized nitrogen and preventing leaching during periods of rain.

Manage the Period between Crops

Organic debris produced by harvesting is easily mineralized into leachable N. Steps to reduce leachable N includes planting of "green manure" crops and delaying ploughing of straw, roots and leaves into the soil.

Rational Irrigation

Poor irrigation has one of the worst impacts on water quality, whereas precision irrigation is one of the least polluting practices as well as reducing net cost of supplied water.

Optimize other Cultivation Techniques

Highest yields with minimum water quality impacts require optimization of practices such as weed, pest and disease control, liming, balanced mineral fertilizers including trace elements, etc.

Agricultural Planning

Implement erosion control techniques that complement topographic and soil conditions.

Organic Fertilizers

Voluntary and legislated control measures in Europe are intended to have the following benefits:

- Reduce the leaching of nutrients.
- Reducing emissions of ammonia.
- Reducing contamination by heavy metals.

The nature of these measures varies by country; however FAO/ECE have summarized the types of voluntary and mandated control as:

- Maximum numbers of animals per hectare based on amount of manure that can be safely applied per hectare of land.
- Maximum quantities of manure that can be applied on the land is fixed, based on the N and P content of the manure.
- Holdings wishing to keep more than a given number of animals must obtain a license.
- The periods during which it is allowed to apply manure to the land have been limited and it is obligatory to work it into the ground immediately afterwards.
- Establishment of regulations on minimum capacity for manure storage facilities.
- Establish fertilizer plans.
- Levies (taxes) on surplus manure.
- Areas under autumn/winter green cover were extended and green fallowing is being promoted.
- Maximum amounts established for spreading of sewage sludge on land based on heavy metal content.
- Change in composition of feed to reduce amount of nutrients and heavy metals.
- Research and implementation of means of reducing ammonia loss.

Sludge Management

Sludge is mentioned here only insofar as the spreading of sludge from municipal wastewater treatment facilities on agricultural land is one method used to get rid of municipal sludge in a way that is perceived to be beneficial. The alternatives are incineration and land fill. FAO/ECE include sludge within the category of organic fertilizers but note that sludge often contains unacceptable levels of heavy metals. Pollution of water by sludge run-off is otherwise the same as for manure.

ECONOMICS OF CONTROL OF FERTILIZER RUN-OFF

Nutrient loss is closely associated with rainfall-run-off events. For phosphorus, which tends to be associated with the solid phase (sediment), run-off losses are directly linked to erosion. Therefore the economics of nutrient control tend to be closely tied to the costs of controlling run-off and erosion. Therefore, this will be treated briefly here. In particular, it is useful to examine the economic cost of nutrient run-off which must be replaced by fertilizers if the land is to remain productive.

The link between erosion, increasing fertilizer application and loss of soil productivity is very direct in many countries. In the Brazilian state of Paraná where agriculture is the base of the state economy, Paraná produces 22 per cent of the national grain production on only 2.4 per cent of the Brazilian territory. Agricultural expansion in Paraná occurred mainly in the period 1950-1970 and was "characterized by short-term agricultural systems leading to continuous and progressive environmental degradation as a result of economic policies and a totally inappropriate land parcelling and marketing system...". Erosion has led to extensive loss of top soil, large-scale gullying and silting of ditches and rivers. The use of fertilizers has risen as a consequence, up 575 per cent over the period 1970-1986 and without any gain in crop yields. Loss of N-P-K from an average erosion of 20 t/ha/yr represents an annual economic loss of US$242 million in nutrients.

Analysis by Elwell and Stocking (1982) of nutrient loss arising from erosion in Zimbabwe shows similar significant economic losses in African situations. Stocking, applying data collected in the 1960s by Hudson to the soil use map of Zimbabwe, calculated an annual loss of 10 million tonnes of nitrogen and 5 million tonnes of phosphorus annually as a consequence of erosion. Roose also cites losses of 98 kg/ha/yr of nitrogen, 29 kg/ha/yr of phosphorus, 39 kg/ha/yr of lime and 39 kg/ha/yr of magnesium from soils of lower Côte d'Ivoire as a result of erosion. This loss is so severe that compensation requires 7 tonnes of fresh manure annually, plus 470 kg of ammonium sulphate, 160 kg of superphosphate, 200 kg dolomite and 60 kg of potassium chloride per hectare per year. Roose notes that it is not surprising that the soil is exhausted after only two years of traditional agriculture. Furthermore, these losses do not take into account additional losses due to harvesting and run-off. Roose summarizes by stating that action against soil erosion is essential in order to manage what he describes as a "terrible" chemical imbalance in soils caused by soil erosion.

Economic losses tend to be higher in tropical countries where soils, rainfall and agricultural practices are more conducive to erosion and reported rates of erosion are much above average.

The World Bank (1992) reported that extrapolation from test-plots of impacts of soil loss on agricultural productivity, indicates some 0.5-1.5 per cent loss of GDP annually for countries such as Costa Rica, Malawi, Mali and Mexico. These losses do not include offsite costs such as reservoir infilling, river sedimentation, damage to irrigation systems, etc.

Soil fertility is a complex issue and nutrient loss is not necessarily nor always a consequence of erosion. Erosion and soil loss is the end member of a variety of physical, vegetative and nutrient factors that lead to soil degradation. Global patterns of fertilizer application, as reported by Joly (1993), indicate however that rapidly rising levels of fertilizer utilization are required merely to maintain soil productivity from a variety of types of loss, including losses due to erosion and, more generally, to soil degradation. In a study of 17 agricultural sub-watersheds in the Lake Balaton district of Hungary, Jolankai

(1986) measured and modelled N and P run-off from a variety of agricultural land uses. He calculated that a selection of control measures (mainly erosion control) would reduce phosphorus loss by 52.8 per cent at a cost of US$ 2500 per ha in remediation measures (in 1986).

AQUACULTURE

Aquaculture is a special case of agricultural pollution. There are two main forms: land-based and water-based systems. Effluent controls are possible on land-based systems, however water-based systems present particular problems. Aquaculture is rapidly expanding in most parts of the developed and developing world, both in freshwater and marine environments. In contrast, coastal fisheries in most countries are declining.

The environmental impact is primarily a function of feed composition and feed conversion (faecal wastes), plus assorted chemicals used as biocides, disinfectants, medicines, etc. Wastage of feed (feed not taken up by the fish) is estimated to be 20 per cent in European aquaculture. Waste feed and faecal production both add substantial nutrient loadings to aquatic systems.

Additional environmental problems include risk of disease and disease transfer to wild fish, introduction of exotic species, impacts on benthic communities and on the eutrophication of water, interbreeding of escaped cultured fish with wild fish with consequent genetic change in the wild population.

Traditional integrated aquaculture systems, as in China, where sewage-fish culture is practised, can be a stabilizing influence in the entire ecosystem. This is recommended, especially in developing countries where water and resources are scarce or expensive.

PROBLEMS OF RESTORATION OF EUTROPHIC LAKES

Eutrophic and hypertrophic lakes tend to be shallow and suffer from high rates of nutrient loadings from point and non-point sources. In areas of rich soils such as the Canadian prairies, lake bottom sediments are comprised of nutrient-enriched soil particles eroded from surrounding soils.

The association of phosphorus with sediment is a serious problem in the restoration of shallow, enriched lakes. P-enriched particles settle to the bottom of the lake and form a large pool of nutrient in the bottom sediments that is readily available to rooted plants and which is released from bottom sediments under conditions of anoxia into the overlying water column and which is quickly utilized by algae. This phosphorus pool, known as the "internal load" of phosphorus, can greatly offset any measures taken by river basin managers to control lake eutrophication by control of external phosphorus sources from agriculture and from point sources. Historically, dredging of bottom sediments was considered the only means of remediating nutrient-rich lake sediments, however, modern technology now provides alternative and more cost-effective methods of controlling internal loads of phosphorus by oxygenation and by chemically treating sediments *in situ* to immobilize the phosphorus. Nevertheless, lake restoration is expensive and must be part of a comprehensive river basin management programme.

PESTICIDES AS WATER POLLUTANTS

The term "pesticide" is a composite term that includes all chemicals that are used to kill or control pests. In agriculture, this includes herbicides (weeds), insecticides (insects), fungicides (fungi), nematocides (nematodes) and rodenticides (vertebrate poisons).

A fundamental contributor to the Green Revolution has been the development and application of pesticides for the control of a wide variety of insectivorous and herbaceous pests that would otherwise diminish the quantity and quality of food produce.

The use of pesticides coincides with the "chemical age" which has transformed society since the 1950s. In areas where intensive monoculture is practised, pesticides were used as a standard method for pest control. Unfortunately, with the benefits of chemistry have also come disbenefits, some so serious that they now threaten the long-term survival of major ecosystems by disruption of predator-prey relationships and loss of biodiversity. Also, pesticides can have significant human health consequences.

While agricultural use of chemicals is restricted to a limited number of compounds, agriculture is one of the few activities where chemicals are intentionally released into the environment because they kill things.

Agricultural use of pesticides is a subset of the larger spectrum of industrial chemicals used in modern society. The American Chemical Society database indicates that there were some 13 million chemicals identified in 1993 with some 500 000 new compounds being added annually. In the Great Lakes of North America, for example, the International Joint Commission has estimated that there are more than 200 chemicals of concern in water and sediments of the Great Lakes ecosystem.

Because the environmental burden of toxic chemicals includes both agriculture and non-agricultural compounds, it is difficult to separate the ecological and human health effects of pesticides from those of industrial compounds that are intentionally or accidentally released into the environment. However, there is overwhelming evidence that agricultural use of pesticides has a major impact on water quality and leads to serious environmental consequences.

Irrigated agriculture, especially in tropical and subtropical environments, usually requires modification of the hydrological regime which, in turn, creates habitat that is conducive to breeding of insects such as mosquitoes which are responsible for a variety of vector-borne diseases. In addition to pesticides used in the normal course of irrigated agriculture, control of vector-borne diseases may require additional application of insecticides such as DDT which have serious and widespread ecological consequences. In order to address this problem, environmental management methods to control breeding of disease vectors are being developed and tested in many irrigation projects.

Historical Development of Pesticides

The history of pesticide development and use is the key to understanding how and why pesticides have been an environmental threat to aquatic systems and why this threat is diminishing in developed countries and remains a problem in many developing countries.

North-south Dilemma over Pesticide Economics

The general progression of pesticide development has moved from highly toxic, persistent and bioaccumulating pesticides such as DDT, to pesticides that degrade rapidly in the environment and are less toxic to non-target organisms. The developed countries have banned many of the older pesticides due to potential toxic effects to man and/or their impacts on ecosystems, in favour of more modern pesticide formulations. In the developing countries, some of the older pesticides remain the cheapest to produce and, for some purposes, remain highly effective as, for example, the use of DDT for malaria control. Developing countries maintain that they cannot afford, for reasons of cost and/ or efficacy, to ban certain older pesticides. The dilemma of cost/efficacy versus ecological impacts, including long range impacts via atmospheric transport and access to modern pesticide formulations at low cost remains a contentious global issue.

In addition to ecological impacts in countries of application, pesticides that have been long banned in developed countries (such as DDT, toxaphene, etc.), are consistently found in remote areas such as the high arctic. Chemicals that are applied in tropical and subtropical countries are transported over long distances by global circulation. The global situation has deteriorated to the point where many countries are calling for a global convention on "POPs" (Persistent Organic Pollutants) which are mainly chlorinated compounds that exhibit high levels of toxicity, are persistent and bioaccumulate. The list is not yet fixed; however, "candidate" substances include several pesticides that are used extensively in developing countries.

FATE AND EFFECTS OF PESTICIDES

Human Health Effects of Pesticides

Perhaps the largest regional example of pesticide contamination and human health is that of the Aral Sea region. UNEP (1993) linked the effects of pesticides to "the level of oncological (cancer), pulmonary and haematological morbidity, as well as on inborn deformities... and immune system deficiencies".

Human health effects are caused by:

- *Skin contact:* Handling of pesticide products.
- *Inhalation:* Breathing of dust or spray.
- *Ingestion:* Pesticides consumed as a contaminant on/in food or in water.

Farm workers have special risks associated with inhalation and skin contact during preparation and application of pesticides to crops. However, for the majority of the population, a principal vector is through ingestion of food that is contaminated by pesticides. Degradation of water quality by pesticide run-off has two principal human health impacts.

The first is the consumption of fish and shellfish that are contaminated by pesticides; this can be a particular problem for subsistence fish economies that lie downstream of major agricultural areas. The second is the direct consumption of pesticide-contaminated water. WHO (1993) has established drinking water guidelines for 33 pesticides. Many health and environmental protection agencies have established "acceptable daily intake" (ADI) values which indicate the maximum allowable daily ingestion over a person's lifetime without appreciable risk to the individual.

Ecological Effects of Pesticides

Pesticides are included in a broad range of organic micro pollutants that have ecological impacts. Different categories of pesticides have different types of effects on living organisms, therefore generalization is difficult. Although terrestrial impacts by pesticides do occur, the principal pathway that causes ecological impacts is that of water contaminated by pesticide run-off. The two principal mechanisms are bioconcentration and biomagnification.

Bioconcentration

This is the movement of a chemical from the surrounding medium into an organism. The primary "sink" for some pesticides is fatty tissue ("lipids"). Some pesticides, such as DDT, are "lipophilic", meaning that they are soluble in and accumulate in, fatty tissue such as edible fish tissue and human fatty tissue. Other pesticides such as glyphosate are metabolized and excreted.

Biomagnification

This term describes the increasing concentration of a chemical as food energy is transformed within the food chain. As smaller organisms are eaten by larger organisms, the concentration of pesticides and other chemicals are increasingly magnified in tissue and other organs. Very high concentrations can be observed in top predators, including man.

The ecological effects of pesticides (and other organic contaminants) are varied and are often inter-related. Effects at the organism or ecological level are usually considered to be an early warning indicator of potential human health impacts.

The major types of effects are listed below and will vary depending on the organism under investigation and the type of pesticide.

Different pesticides have markedly different effects on aquatic life which makes generalization very difficult. The important point is that many of these effects are chronic (not lethal), are often not noticed by casual observers, yet have consequences for the entire food chain.

- Death of the organism.
- Cancers, tumours and lesions on fish and animals.
- Reproductive inhibition or failure.
- Suppression of immune system.
- Disruption of endocrine (hormonal) system
- Cellular and DNA damage.
- Teratogenic effects (physical deformities such as hooked beaks on birds).
- Poor fish health marked by low red to white blood cell ratio, excessive slime on fish scales and gills, etc.
- Intergenerational effects (effects are not apparent until subsequent generations of the organism).
- Other physiological effects such as egg shell thinning.

These effects are not necessarily caused solely by exposure to pesticides or other organic contaminants, but may be associated with a combination of environmental stresses such as eutrophication and pathogens. These associated stresses need not be large to have a synergistic effect with organic micro pollutants.

Ecological effects of pesticides extend beyond individual organisms and can extend to ecosystems.

Swedish work indicates that application of pesticides is thought to be one of the most significant factors affecting biodiversity. Jonsson report that the continued decline of the

Swedish partridge population is linked to changes in land use and the use of chemical weed control. Chemical weed control has the effect of reducing habitat, decreasing the number of weed species and of shifting the balance of species in the plant community.

Swedish studies also show the impact of pesticides on soil fertility, including inhibition of nitrification with concomitant reduced uptake of nitrogen by plants.

These studies also suggest that pesticides adversely affect soil micro-organisms which are responsible for microbial degradation of plant matter (and of some pesticides) and for soil structure.

NATURAL FACTORS THAT DEGRADE PESTICIDES

In addition to chemical and photochemical reactions, there are two principal biological mechanisms that cause degradation of pesticides.

These are:

- Microbiological processes in soils and water.
- Metabolism of pesticides that are ingested by organisms as part of their food supply.

While both processes are beneficial in the sense that pesticide toxicity is reduced, metabolic processes do cause adverse effects in, for example, fish. Energy used to metabolize pesticides and other xenobiotics (foreign chemicals) is not available for other body functions and can seriously impair growth and reproduction of the organism.

Degradation of Pesticides in Soil

"Many pesticides dissipate rapidly in soils. This process is mineralization and results in the conversion of the pesticide into simpler compounds such H_2O, CO_2 and NH_3.

While some of this process is a result of chemical reactions such as hydrolysis and photolysis, microbiological catabolism and metabolism is usually the major route of mineralization. Soil micro biota utilize the pesticide as a source of carbon or other nutrients. Some chemicals (for example 2,4-D) are quite rapidly broken down in soil while others are less easily attacked (2,4,5-T). Some chemicals are very persistent and are only slowly broken down (atrazine)".

Process of Metabolism

Metabolism of pesticides in animals is an important mechanism by which organisms protect themselves from the toxic effects of xenobiotics (foreign chemicals) in their food supply. In the organism, the chemical is transformed into a less toxic form and either excreted or stored in the organism.

Different organs, especially the liver, may be involved, depending on the chemical. Enzymes play an important role in the metabolic process and the presence of certain enzymes, especially "mixed" function oxygenases (MFOs) in liver, is now used as an indicator that the organism has been exposed to foreign chemicals.

AGRICULTURAL USE OF SEWAGE SLUDGE

Sewage sludge has valuable agronomic properties in agriculture. In using sewage sludge account must be taken of the nutrient needs of the plants without, however, impairing neither the quality of the soil nor that of surface and ground water. Some heavy metals present in sewage sludge may be toxic to plants and humans. Sewage sludge may be used in agriculture provided that the Member State concerned regulates its use.

The Directive lays down limit values for concentrations of heavy metals in the soil, in sludge and for the maximum annual quantities of heavy metals which may be introduced into the soil.

The use of sewage sludge is prohibited if the concentration of one or more heavy metals in the soil exceeds the limit values laid down in accordance.

The Member States must therefore take the necessary steps to ensure that those limit values are not exceeded as a result of using sludge. Sludge must be treated before being used in agriculture but the Member States may authorise the use of untreated sludge if it is injected or worked into the soil.

CHARACTERISTICS OF SEWAGE SLUDGE

Most wastewater treatment processes produce a sludge which has to be disposed of. Conventional secondary sewage treatment plants typically generate a primary sludge in the primary sedimentation stage of treatment and a secondary, biological, sludge in final sedimentation after the biological process. The characteristics of the secondary sludge vary with the type of biological process and, often, it is mixed with primary sludge before treatment and disposal.

Approximately one half of the costs of operating secondary sewage treatment plants in Europe can be associated with sludge treatment and disposal.

Land application of raw or treated sewage sludge can reduce significantly the sludge disposal cost component of sewage treatment as well as providing a large part of the nitrogen and phosphorus requirements of many crops. Very rarely do urban sewerage systems transport only domestic sewage to treatment plants; industrial effluents and storm-water run-off from roads and other paved areas are frequently discharged into sewers.

Thus sewage sludge will contain, in addition to organic waste material, traces of many pollutants used in our modern society. Some of these substances can be phytotoxic and some toxic to humans and/or animals so it is necessary to control the concentrations in the soil of potentially toxic elements and their rate of application to the soil.

The risk to health of chemicals in sewage sludge applied to land has been reviewed by Dean and Suess.

Sewage sludge also contains pathogenic bacteria, viruses and protozoa along with other parasitic helminths which can give rise to potential hazards to the health of humans, animals and plants. A WHO Report on the risk to health of microbes in sewage sludge applied to land identified salmonellae and *Taenia* as giving rise to greatest concern.

The numbers of pathogenic and parasitic organisms in sludge can be significantly reduced before application to the land by appropriate sludge treatment and the potential health risk is further reduced by the effects of climate, soil-microorganisms and time after the sludge is applied to the soil. Nevertheless, in the case of certain crops, limitations on planting, grazing and harvesting are necessary.

Apart from those components of concern, sewage sludge also contains useful concentrations of nitrogen, phosphorus and organic matter.

The availability of the phosphorus content in the year of application is about 50 per cent and is independent of any prior sludge treatment. Nitrogen availability is more dependent on sludge treatment, untreated liquid sludge and dewatered treated sludge releasing nitrogen slowly with the benefits to crops being realised over a relatively long period.

Liquid anaerobically-digested sludge has high ammonia-nitrogen content which is readily available to plants and can be of particular benefit to grassland.

The organic matter in sludge can improve the water retaining capacity and structure of some soils, especially when applied in the form of dewatered sludge cake.

Directive prohibits the sludge from sewage treatment plants from being used in agriculture unless specified requirements are fulfilled, including the testing of the sludge and the soil. Parameters subject to the provisions of the Directive include the following:

- Dry matter (per cent),
- Organic matter (per cent dry solids),
- Copper (mg/kg dry solids),
- Nickel (mg/kg dry solids),
- pH,
- Nitrogen, total and ammoniacal (per cent dry solids),
- Phosphorus, total (per cent dry solids),
- Zinc (mg/kg dry solids),
- Cadmium (mg/kg dry solids),
- Lead (mg/kg dry solids),
- Mercury (mg/kg dry solids),
- Chromium (mg/kg dry solids).

To these parameters the UK Department of the Environment has added molybdenum, selenium, arsenic and fluoride in the recent 'Code of Practice for Agricultural Use of Sewage Sludge'. Sludge must be analyzed for the Directive parameters at least once every 6 months and every time significant changes occur in the quality of the sewage treated.

The frequency of analysis for the additional four parameters may be reduced to not less than once in five years provided that their concentrations in the sludge are consistently no greater than the following reference concentrations: Mb - 3mg/kg dry solids, Se - 2mg/kg dry solids, As -2mg/kg dry solids and Fl - 200mg/kg dry solids.

SLUDGE TREATMENT

Except when it is to be injected or otherwise worked into the soil, sewage sludge should be subjected to biological, chemical or thermal treatment, long-term storage or other appropriate process designed to reduce its fermentability and health hazards resulting from its use before being applied in agriculture. Table lists sludge treatment and handling processes which have been used in the UK to achieve these objectives.

The second edition of a 'Manual of Good Practice on Soil Injection of Sewage Sludge' has been produced by the Water Research Centre in the UK and describes suitable equipment and techniques for what is now the only method permissible within the EEC for applying untreated sludges to grassland.

Table: Examples of Effective Sludge Treatment Processes

Process	Descriptions
Sludge Pasteurization	Minimum of 30 minutes at 70°C or minimum of 4 hours at 55° C (or appropriate intermediate conditions), followed in all cases by primary mesophilic anaerobic digestion.
Mesophilic Anaerobic Digestion	Mean retention period of at least 12 days primary digestion in temperature range 35°C +/- 3°C or of at least 20 days primary digestion in temperature range 25°C +/- 3°C followed in each case by a secondary stage which provides a mean retention period of at least 14 days.
Thermophilic Aerobic All Digestion	Mean retention period of at least 7 days digestion. sludge to be subject to a minimum of 55°C for a period of at least 4 hours.
Composting (Windrows or Aerated Piles)	The compost must be maintained at 40°C for at least 5 days and for 4 hours during this period at a minimum of 55°C within the body of the pile followed by a period of maturation adequate to ensure that the compost reaction is substantially complete.
Lime Stabilization of Liquid Sludge	Addition of lime to raise pH to greater than 12.0 and sufficient to ensure that the pH is not less than 12 for a minimum period of 2 hours. The sludge can t then be used directly
Liquid Storage	Storage of untreated liquid sludge for a minimum period of 3 months.
Dewatering and Storage	Conditioning of untreated sludge with lime or other coagulants followed by dewatering and storage of the cake for a minimum period of 3 months if sludge has been subject to primary mesophilic anaerobic digestion, storage to be for a minimum period of 14 days.

SLUDGE APPLICATION

The concentrations of potentially toxic elements in arable soils must not exceed certain prudent limits within the normal depth of cultivation as a result of sludge application.

No sludge should be applied at any site where the soil concentration of any of the parameters with the exception of molybdenum, exceed these limits. Maximum permissible concentrations of the potentially toxic elements in soil after application of sewage sludge.

For zinc, copper and nickel, the maximum permissible concentrations vary with the pH of the soil because it is known that crop damage from phytotoxic elements is more likely to occur on acid soils. When sludge is applied to the surface of grassland, the concentrations of potentially toxic elements should be determined in soil samples taken to a depth of 7.5 cm.

In order to minimize injestion of lead, cadmium and fluoride by livestock, the addition of these elements through sludge application to the surface should not exceed 3 times the 10 year average annual rates.

EFFECTS OF SLUDGE ON SOILS AND CROPS

The natural background concentration of metals in the soil is normally less available for crop uptake and hence less hazardous than metals introduced through sewage sludge applications.

Research carried out in the U.K has shown that the amounts of Cd, Ni, Cu, Zn and Pb applied in liquid sludge at three experimental sites could be accounted for by soil profile analyses five years after sludge applications, with the exception of Cu and Zn applied to a calcareous loam soil.

These field experiments also determined the extent of transfer of metals from sludge-treated soil into the leaves and edible parts of six crops of major importance to UK agriculture and the effect of metals on yields of these crops.

Although all the plots received sufficient inorganic fertilizer to meet crop requirements for nutrients, the applications of sludge had some effects on crop yields.

In 60 per cent of the cases studied crop yields were not significantly affected but in 26 per cent of the cases liquid sludge application resulted in significantly increased crop yields, attributed to the beneficial effects on soil structure.

Reductions in wheat grain yield, from 6 - 10 per cent, were noted on the clay and calcareous loam soils treated with liquid sludge and the sandy loam and clay soils treated with bed-dried sludge.

However, this yield reduction was not thought to be due to metals but the most likely explanation was lodging of the crop as a result of excessive nitrogen in the soil.

Increases in metal concentrations in the soil due to sludge applications produced significant increases in Cd, Ni, Cu and Zn concentrations in the edible portion of most of the crops grown: wheat, potato, lettuce, red beet, cabbage and ryegrass.

In most cases there was no significant increase of Pb in crop tissue in relation to Pb in the soil from sludge application, suggesting that lead is relatively unavailable to crops from the soil. The availability of metals to crops was found to be lower in soil treated with bed-dried sludge cake compared with liquid sludge, the extent being dependent on the crop.

Even though the Ni, Cu and Zn concentrations in the soils treated with high rates of application of liquid and bed-dried sludges were close to the maximum levels set out in the EC Directive and the zinc equivalent of sludge addition exceeded the maximum permitted in U.K. guidelines, no phytotoxic effects of metals were evident, with one exception. This was in lettuce grown on clay soil, when Cu and Zn levels exceeded upper critical concentrations at high rates of sludge application.

PLANTING, GRAZING AND HARVESTING CONSTRAINTS

To minimize the potential risk to the health of humans, animals and plants it is necessary to coordinate sludge applications in time with planting, grazing or harvesting operations. Sludge must not be applied to growing soft fruit or vegetable crops nor used where crops are grown under permanent glass or plastic structures.

The EC Directive requires a mandatory 3-week no grazing period for treated sludge applied to grassland but prohibits the spreading of untreated sludge on grassland unless injected. -

Treated sludge can be applied to growing cereal crops without constraint but should not be applied to growing turf within 3 months of harvesting or to fruit trees within 10 months of harvesting.

When treated sludge is applied before planting such crops as cereals, grass, fodder, sugar beet, fruit trees, etc., no constraints apply but in the case of soft fruit and vegetables, the treated sludge should not be applied within 10 months of crop harvesting. In general, untreated sludge should only be cultivated or injected into the soil before planting crops but can be injected into growing grass or turf, with the constraints on minimum time to harvesting as already mentioned.

ENVIRONMENTAL PROTECTION

Care should always be taken when applying sewage sludge to land to prevent any form of adverse environmental impact. The sludge must not contain non-degradable materials, such as plastics, which would make land disposal unsightly. Movement of sludge by tanker from sewage treatment plant to agricultural land can create traffic problems and give rise to noise and odour nuisance.

Vehicles should be carefully selected for their local suitability and routes chosen so as to minimize inconvenience to the public. Access to fields should be selected after consultation with the highway authority and special care must be taken to prevent vehicles carrying mud onto the highway.

Ocour control is the most important environmental dimension of sludge application to land. Enclosed tankers should be used for transporting treated sludge, which tends to be less odorous than raw sludge.

Discharge points for sludge from tankers or irrigators should be as near to the ground as is practicable and the liquid sludge trajectory should be kept low so as to minimize spray drift and visual impact. Untreated sludge should be injected under the soil surface using special vehicles or tankers fitted with injection equipment.

Great care is needed to prevent sludge running off onto roads or adjacent land, depending on topography, soil and weather conditions.

On sloping land there is the risk of such run-off reaching watercourses and causing serious water pollution. Sludge application rates must be adjusted accordingly and, under certain circumstances, spreading might have to be discontinued.

In addition to the problem of surface run-off, pollution may arise from the percolation of liquid sludge into land drains, particularly when injection techniques are used or liquid sludge is applied to dry fissured soils.

In highly sensitive water pollution areas, sludge should be used only in accordance with the requirements of the pollution control authority as well as of good farming practice. Sludge storage on farms can optimize the transport and application operations but every effort must be made to ensure that storage facilities are secure.

6

Hydrology and Water Resources

HYDROLOGY IS THE STUDY OF WATER

Water is one of our most important natural resources. Without it, there would be no life on earth. The supply of water available for our use is limited by nature. Although there is plenty of water on earth, it is not always in the right place, at the right time and of the right quality. Adding to the problem is the increasing evidence that chemical wastes improperly discarded yesterday are showing up in our water supplies today. Hydrology has evolved as a science in response to the need to understand the complex water systems of the Earth and help solve water problems. Hydrologists play a vital role in finding solutions to water problems, and interesting and challenging careers are available to those who choose to study hydrology.

WATER AND PEOPLE

Estimates of water use in the United States indicate that about 408 billion gallons per day (one thousand million gallons per day, abbreviated Bgal/d) were withdrawn for all uses during 2000. This total has varied less than 3 per cent since 1985 as withdrawals have stabilized for the two largest uses—thermoelectric power and irrigation.

Fresh ground-water withdrawals (83.3 Bgal/d) during 2000 were 14 per cent more than during 1985. Fresh surface-water withdrawals for 2000 were 262 Bgal/d, varying less than 2 per cent since 1985. Much of our water use is hidden. Think about what you had for lunch.

A hamburger, for example, requires water to raise wheat for the bun, to grow hay and corn to feed the cattle and to process the bread and beef. Together with french fries and a soft drink, this all-American meal uses about 1,500 gallons of water--enough to fill a small swimming pool.

How about your clothes? To grow cotton for a pair of jeans takes about 400 gallons. A shirt requires about 400 gallons. How do you get to school or to the store? To produce the amount of finished steel in a car has in the past required about 32,000 gallons of water. Similarly, the steel in a 30-pound bicycle required

480 gallons. This shows that industry must continue to strive to reduce water use through manufacturing processes that use less water, and through recycling of water.

WHAT IS HYDROLOGY?

Hydrology is the science that encompasses the occurrence, distribution, movement and properties of the waters of the earth and their relationship with the environment within each phase of the hydrologic cycle.

The water cycle, or hydrologic cycle, is a continuous process by which water is purified by evaporation and transported from the earth's surface (including the oceans) to the atmosphere and back to the land and oceans.

All of the physical, chemical and biological processes involving water as it travels its various paths in the atmosphere, over and beneath the earth's surface and through growing plants, are of interest to those who study the hydrologic cycle. There are many pathways the water may take in its continuous cycle of falling as rainfall or snowfall and returning to the atmosphere.

It may be captured for millions of years in polar ice caps. It may flow to rivers and finally to the sea. It may soak into the soil to be evaporated directly from the soil surface as it dries or be transpired by growing plants. It may percolate through the soil to ground water reservoirs (aquifers) to be stored or it may flow to wells or springs or back to streams by seepage. They cycle for water may be short, or it may take millions of years. People tap the water cycle for their own uses. Water is diverted temporarily from one part of the cycle by pumping it from the ground or drawing it from a river or lake. It is used for a variety of activities such as households, businesses and industries; for irrigation of farms and parklands; and for production of electric power. After use, water is returned to another part of the cycle: perhaps discharged downstream or allowed to soak into the ground. Used water normally is lower in quality, even after treatment, which often poses a problem for downstream users. The hydrologist studies the fundamental transport processes to be able to describe the quantity and quality of water as it moves through the cycle (evaporation, precipitation, streamflow, infiltration, ground water flow, and other components). The engineering hydrologist, or water resources engineer, is involved in the planning, analysis, design, construction and operation of projects for the control, utilization, and management of water resources. Water resources problems are also the concern of meteorologists, oceanographers, geologists, chemists, physicists, biologists, economists, political scientists, specialists in applied mathematics and computer science, and engineers in several fields.

WHAT HYDROLOGISTS DO?

Hydrologists apply scientific knowledge and mathematical principles to solve water-related problems in society: problems of quantity, quality and availability. They may be concerned with finding water supplies for cities or irrigated farms, or controlling river flooding or soil erosion. Or, they may work in environmental protection: preventing or cleaning up pollution or locating sites for safe disposal of hazardous wastes.

Persons trained in hydrology may have a wide variety of job titles. Scientists and engineers in hydrology may be involved in both field investigations and office work. In the field, they may collect basic data, oversee testing of water quality, direct field crews and work with equipment. Many jobs require travel, some abroad. A hydrologist may spend considerable time doing field work in remote and rugged terrain. In the office, hydrologists do many things such as interpreting hydrologic data and performing analyses for determining possible water supplies. Much of their work relies on computers for organizing, summarizing and analysing masses of data, and for modeling studies such as the prediction of flooding and the consequences of reservoir releases or the effect of leaking underground oil storage tanks. The work of hydrologists is as varied as the uses of water and may range from planning multimillion dollar interstate water projects to advising homeowners about backyard drainage problems.

HYDROLOGY AND HYDROGEOLOGY

Hydrologists and hydrogeologists are water scientists who study the properties of freshwater and its distribution on the continents. (Oceanographers study the physical and chemical properties of salt water in the oceans.) Together, hydrology and hydrogeology provide information on how to manage and protect freshwater, humans most essential natural resource. Hydrology and hydrogeology are distinct fields of study that employ different methods and techniques, but they overlap to provide a complete picture of Earth's freshwater resources.

Hydrology is a branch of engineering that deals with the physical properties of surface freshwater, such as lakes and rivers, and with its chemical interactions with other substances. Hydrogeology is a subfield of geology (study of Earth) that, by definition, specifically addresses groundwater-water moving through tiny openings in rock and soil layers beneath the land surface. In practice, ground and surface water interact as a single system. Surface water seeps into the ground and groundwater emerges to the surface. Hydrogeologists work to explain the geological effects of surface water in rivers, streams and lakes, and hydrologists lend their technical expertise to the mechanics and chemistry of moving groundwater.

HYDROLOGY

Hydrologists use mathematics and experimental techniques to determine water's general properties, to make specific observations of freshwater environments, and to design water management systems that contain and direct water. Britain's Centre for Hydrology, a government environmental research agency, describes its mission as an effort to answer two questions about the Earth's freshwater: Why is the natural environment as it is? and What is it likely to be in the future?

Today, humans, not natural processes, manage the flow, distribution, and allocation of almost all of Earth's surface waters. Dams, levees, and reservoirs (natural or man-made lakes) direct and contain the water of the world's largest and most heavily used river systems: the Nile, Yangtze, Amazon, Ganges, and Mississippi. U.S. Army Corps of Engineers hydrologists, along with their counterparts at the Bureau of Reclamation (which has jurisdiction over rivers west of the Rockies), control the flow and distribution of all the surface waters of the United States. Hydrological studies generally seek to understand how water moves between and through bodies of freshwater, such as lakes, streams, aquifers (water–bearing soil and rock layers), and reservoirs, and even the atmosphere (mass of air surrounding Earth) over time. Hydrologists usually begin with data from a particular region. They collect numerous measurements of water system conditions such as rainfall totals, lake and reservoir levels, river discharges (the volume of water that flows through a river in a given time), and current speeds, air and water temperatures, and humidity (amount of water vapour in the atmosphere) at specific sites. Then they merge the data into a computerized model that shows how water has moved through the region's various freshwater reservoirs in the past.

Once a hydrologist has constructed a model of a watershed (land area that contributes water to a stream, lake, or aquifer) or reservoir (body of water), the hydrologist tests it by comparing its calculated predictions with actual results in the natural environment. The model is adjusted to better match the real world. The more times the scientist repeats the process of adding new data to the model and retesting it, the more accurate it becomes. Computer models help hydrologists predict the ways that natural and man-made changes like droughts (uncommonly dry weather), heavy snows and rains, and new dams across rivers affect water supplies and flows in the future. Many models also include information about the way water physically and chemically interacts with rocks and minerals on its path through the system.

These models can be used to track contaminants and predict water quality changes over time. One example of how hydrologists' work can affect a regions water supply is in the city of Denver, Colorado when the city had very low reserves of freshwater in its reservoirs. There had been lower than average snowfall in the nearby Rocky Mountains and the city's water supplies ran low without their usual influx of spring melt water.

Hydrologists at the U.S. Geological Survey in Denver used data and computer models to help the city distribute limited water during the drought period, and to plan for future dry periods. Models of Denver's surface water flows and seasonal patterns also help water managers protect and regulate the city's water supply from contamination by such human waste products as agricultural fertilizers and chemicals, industrial wastes, and sewage.

HYDROGEOLOGY

Hydrogeologists are concerned mostly with groundwater and how geologic features affect groundwater storage, flow, and replenishment. Like other geologists, they use observations of rock types and geologic structures on the land surface together with subsurface samples to map folded, faulted (broken), and fractured (cracked) rock layers and bodies beneath the land surface. Hydrogeologic maps show the locations and shapes of aquifers, and the distribution of less permeable (leaky) layers called aquicludes that confine and pressurize groundwater within aquifers.

They identify systems of caves, cavities, and fractures where groundwater may flow more quickly along specific routes. They also show the areas where surface water enters aquifers (recharge zones), and places where groundwater reemerges at springs and seeps (discharge zones).

Once a hydrogeologist has mapped out the dimensions, physical characteristics, and "plumbing" of a groundwater reservoir, he or she sets out to understand how water moves through the system. Measurements of flows at recharge and discharge points show the rates at which water is entering and leaving the aquifer as well as the time an average water molecule (smallest part of water that has its properties) takes to travel through the system (residence time.) Water levels and pressures in wells indicate flow patterns and rates inside the aquifer.

Groundwater can become polluted. Hydrogeologists also use their understanding of groundwater flow patterns to predict how contaminants might enter an aquifer, move through the system, and reemerge in a distant spring or well.

Sometimes an aquifer or soil layer acts as a filter that improves water quality as contaminants move through an aquifer. Others transport polluted water quickly to a discharge site.

Some rock layers even contribute hazardous dissolved chemicals to the groundwater. Hydrogeologists collect water samples and monitor water quality within aquifers. They also conduct laboratory and computer experiments to better understand groundwater's chemical interactions. Cities and regions that depend on groundwater require detailed hydrogeologic maps, and a good understanding of how water moves through their aquifer, to effectively manage groundwater resources.

In many states and countries, groundwater is common, public property. Most rules and regulations regarding groundwater consumption and contamination were written in an era when groundwater systems were poorly understood and were considered unending sources of clean freshwater. Today, many regions with large human populations and fragile natural ecosystems (communities of plants and animals) depend on limited, shared groundwater resources. As such, the actions of some of a groundwater reservoir's users can negatively affect the water supply and quality for other people, as well as for the aquifer's plants and animals.

HYDROPOWER

Hydropower is energy that is generated by moving water. Today, hydropower facilities make electricity by converting kinetic (moving) energy into mechanical (machine) energy as water flows in a river or over a dam. Electricity made at hydropower facilities can be carried away, via power transmission lines, and sold to homes and businesses. Hydropower is a relatively inexpensive, non–polluting form of renewable energy. Canada and the United States are currently the world's top hydroelectric producers. Other countries that use hydropower on a large scale include Brazil, China, Russia, Norway, Japan, India, Iceland, Sweden, and France.

Hydropower produces about 10 per cent of United State's electricity, in contrast to Norway, who generates nearly 99 per cent of its electricity from hydropower. Hydropower is used nationwide, but is primarily used in the western coastal United States where other energy resources such as coal are limited. Hydropower is important to the United States economy because it supplies electricity to a growing population and industry.

HYDROPOWER IN HISTORY

Humans have harnessed water power for thousands of years, using the mechanical energy of moving water to turn wooden wheels to power mills that sawed lumber and processed grains.

Water either fell onto the wheel and caused it to turn or the wheel was placed in the river and the river's current (a steady flow of water in a prevailing direction) turned the wheel.

The wheel was attached to other levers and gears inside the mill that did the work needed. During the 1700s to 1800s, mechanical hydropower was used extensively in the United States and elsewhere for milling and pumping. It began to be widely used to supply electricity in the late 1800s.

In 1882 the world's first hydropower facility was built on the Fox River in Appleton, Wisconsin. The Fox River facility generated electricity for local industries.

By the 1920s, following the development of the electrical motor and the demand for electricity that followed, hydropower accounted for about 40 per cent of the U.S. electricity supply. Since then, other energy technologies have developed that are less expensive than hydropower. In 1933, President Franklin D. Roosevelt (1882–1945) signed the Tennessee Valley Authority (TVA) Act.

The purpose of the TVA was to construct dams on the Tennessee River that would aid in river navigation, control flooding, provide water for irrigation (watering crops) and drinking, and provide hydropower to Tennessee River Valley residents. In the American West, hydropower aided in the production of the dams themselves, moving and lifting construction materials and providing energy for lights to make round-the-clock work possible. Surplus energy was sold to neighbouring farms and homes, which in turn paid for the operation and building costs of the dams. Hydropower facility development was at its peak during the 1930s and 1940s. The Hoover Dam on the Nevada-Arizona border and the Grand Coulee dam located on the Columbia River in Washington state were constructed during this period. The Grand Coulee dam remains the largest concrete structure in the United States.

HYDROELECTRIC TECHNOLOGY

Electricity is one of the most important types of energy because it allows people to perform the work needed to light homes and power appliances and computers. Hydropower generation utilizes the principles of electromagnetic induction (creating an electrical current, the flow of electricity, by moving a magnet through a wire coil), first described by English chemist and physicist Michael Faraday (1791–1867) in 1831 when he made the first generator. A generator is a machine that converts mechanical energy to electrical energy. Energy is the power to do work. Energy cannot be created or destroyed; it merely changes state, as occurs when electrical energy is harnessed from the mechanical energy of moving water. No matter the type or size of the hydropower facility, electricity is generated in much the same way in each one. The dam or the natural elevation drop in a river creates head, or a certain height over which the water flows as it is released from the dam or as it flows downstream. As water is released from a reservoir in an impoundment or pumped storage facility, or diverted from a river through a control gate, it flows by gravity (the attraction between two masses) to a turbine. A turbine is a device that transforms the energy in moving fluid or wind to rotary mechanical energy.

As the water flows past the turbine blades, the turbine blades spin and turn a rotor (the moving part of an electric generator) much like wind spins a pinwheel. Giant magnets inside the rotor move past coils of copper wire and create an alternating current. An alternating current (AC) is produced when electrons wiggle back and forth between atoms. The used water returns to the river through pipes. The alternating current moves through a series of devices called transformers that can increase the voltage (energy required to move a charge from one point to another, similar to pressure).

The increase in voltage allows the electricity to travel faster and more efficiently through power lines (important when the hydropower facility is in a remote location) from the hydropower facility to a town's electricity facility. At the local facility transformers are used again to reduce the voltage to a level that is safe for the electricity to be used in homes and businesses.

TYPES OF HYDROPOWER FACILITIES

There are three types of hydropower facilities: impoundment, pumped storage, and diversion. Impoundment and pumped storage facilities require dams. In the United States, hydropower is a very small percentage of the primary purpose of dams. Usually a dam is built first for other reasons, such as water storage and flood control, and a hydropower facility is incorporated later if there is a demonstrated need for electricity. Impoundment facilities require a large dam that allows river water to be stored in a reservoir. When water is released from the reservoir, the water flows downward through a penstock (pipe) in the dam and spins turbines, thereby creating mechanical energy that is then used to power electric generators. Transmission lines carry electricity away from the impoundment facility to local distributors who sell the electricity to homes and businesses. Pumped storage hydropower facilities also require dams to operate.

In periods of low electricity demand water is pumped from a lower reservoir to a higher reservoir. When electrical demand increases, water is released from the higher reservoir back into the lower reservoir through a penstock, turning turbines that power electric generators. Pumped storage and impoundment hydropower facilities provide a reliable source of electricity because water flow from reservoirs can be controlled so that electricity production meets demand. Diversion hydropower facilities (sometimes called run-ofriver systems) are smaller than impoundment facilities and do not require a dam. Instead, diversion facilities use a river's natural flow to generate electricity. The amount of electricity produced depends upon the river's rate of flow (volume of water flow within a period of time) and the river's elevation change at the diversion facility site.

In diversion hydropower facilities, river water is channeled through a canal (artificial waterway that controls water flow, in this case, to a turbine) or penstock. Diversion facilities are less reliable than impoundment or pumped storage facilities because flow rates in rivers can change drastically depending on the amount of rainfall or spring meltwater that supplies the river.

SIZES OF HYDROPOWER FACILITIES

The size of a hydropower facility depends upon the amount of energy that can be generated at the facility. Some hydropower plants may produce electricity for only one to a few homes. These facilities are called micro-hydropower plants.

In microhydropower facilities the total change in elevation of the flowing water is only about 100 feet (30 meters). Energy output can be increased, however, with higher water flow. Larger hydroelectric power facilities such as the Hoover power plant in Nevada, however, can provide electricity to more than one million consumers.

BENEFITS AND DRAWBACKS OF HYDROPOWER

Hydropower is an efficient, clean, and reliable source of energy. Once hydropower plants are constructed and the technology is in place, the cost of hydropower is the lowest of all energy sources. No fuel is used during hydropower production; water returns to the river. No pollutants are released into the air during electrical generation, so the air around hydropower plants remains clean. Hydropower facilities can respond to high demands for electricity through reservoir storage, even in times of water shortage.

The negative effects of hydropower generally come from large–scale facilities. Impoundment or pumped storage facilities often alter wildlife habitats along rivers because large dams must be built to provide a reservoir. Dams can flood areas close to a river or lake, obstruct fish migration (periodic movement from one region to another), and affect water quality and flow downstream.

Humans are sometimes displaced from their homes when dams are built. In diversion facilities, seasonal and annual fluctuations (variations) in water supply can negatively affect electrical production if river flow rates become too low. Regulations and licensing permits for dams and hydropower facilities as well as electrical transportation from remote hydropower are often costly.

THE FUTURE OF HYDROPOWER

Recent energy shortages in the United States has spurred the government to study hydropower's future potential.

The Department of Energy has identified 5,677 sites in the United States that have potential for large-scale hydropower development, but many of these sites are not possible because of economic drawback, such as their remote location, environmental issues, and other circumstances. Because a change to hydropower as a primary energy source could help lessen air pollution and reduce demands for fossil fuels (oil and gas), hydropower will likely be further explored.

SURFACE WATER

Most cities meet their needs for water by withdrawing it from the nearest river, lake or reservoir. Hydrologists help cities by collecting and analysing the data needed to predict how much water is available from local supplies and whether it will be sufficient to meet the city's projected future needs. To do this, hydrologists study records of rainfall, snowpack depths and river flows that are collected and compiled by hydrologists in various government agencies. They inventory the extent river flow already is being used by others.

Managing reservoirs can be quite complex, because they generally serve many purposes. Reservoirs increase the reliability of local water supplies. Hydrologists use topographic maps and aerial photographs to determine where the reservoir shorelines will be and to calculate reservoir depths and storage capacity. This work ensures that, even at maximum capacity, no highways, railroads or homes would be flooded.

Deciding how much water to release and how much to store depends upon the time of year, flow predictions for the next several months, and the needs of irrigators and cities as well as downstream water-users that rely on the reservoir. If the reservoir also is used for recreation or for generation of hydroelectric power, those requirements must be considered.

Decisions must be coordinated with other reservoir managers along the river. Hydrologists collect the necessary information, enter it into a computer, and run computer models to predict the results under various operating strategies. On the basis of these studies, reservoir managers can make the best decision for those involved.

The availability of surface water for swimming, drinking, industrial or other uses sometimes is restricted because of pollution. Pollution can be merely an unsightly and inconvenient nuisance, or it can be an invisible, but deadly, threat to the health of people, plants and animals. Hydrologists assist public health officials in monitoring public water supplies to ensure that health standards are met.

When pollution is discovered, environmental engineers work with hydrologists in devising the necessary sampling programme. Water quality in estuaries, streams, rivers and lakes must be monitored, and the health of fish, plants and wildlife along their stretches surveyed. Related work concerns acid rain and its effects on aquatic life, and the behaviour of toxic metals and organic chemicals in aquatic environments. Hydrologic and water quality mathematical models are developed and used by hydrologists for planning and management and predicting water quality effects of changed conditions.

Simple analyses such as pH, turbidity, and oxygen content may be done by hydrologists in the field. Other chemical analyses require more sophisticated laboratory equipment. In the past, municipal and industrial sewage was a major source of pollution for streams and lakes. Such wastes often received only minimal treatment, or raw wastes were dumped into rivers. Today, we are more aware of the consequences of such actions, and billions of dollars must be invested in pollution-control equipment to protect the waters of the earth. Other sources of pollution are more difficult to identify and control. These include road deicing salts, storm run-off from urban areas and farmland, and erosion from construction sites.

SURFACE WATER USE

The Nation's surface-water resources-the water in the nation's rivers, streams, creeks, lakes, and reservoirs—are vitally important to our everyday life. The main uses of surface water include drinking–water and other public uses, irrigation uses, and for use by the thermoelectric–power industry to cool electricity–generating equipment. The majority of water used for thermoelectric power, public supply, irrigation, mining, and industrial purposes came from surface–water sources.

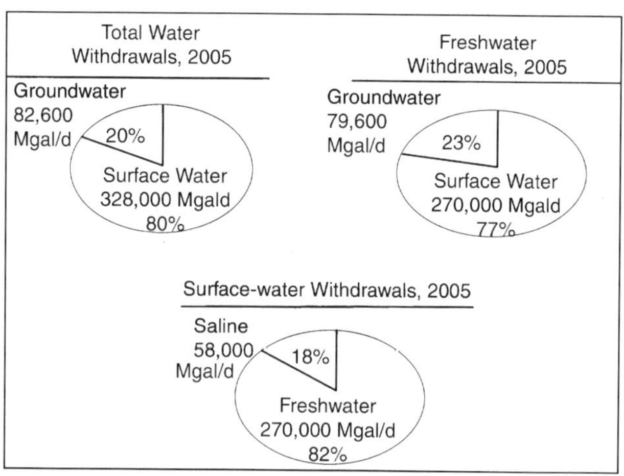

Of all the water used in the United States in 2005, about 80 per cent came from surface–water sources. Water from groundwater sources accounted for the remaining 20 per cent.

Over 85 per cent of all water used in 2005 was freshwater, although saline water was heavily used in the thermoelectric-power industry, and, to a lesser extent, for industrial and mining purposes.

About 77 per cent of the freshwater used in the United States in 2005 came from surface–water sources.

The other 23 per cent came from groundwater. Surface water is an important natural resource used for many purposes, especially irrigation and public supply.

Surface–water Use, by Category of Use, 2005

For 2005, most of the fresh surface–water withdrawals, 53 per cent, was used in the thermoelectric-power industry to cool electricity–generating equipment. Water used in this manner is most often returned to its source.

That is why the more significant use of surface water is irrigation, which used about 28 per cent of all fresh surface water, but, ignoring thermoelectric–power withdrawals, irrigation accounted for about 58 per cent of the Nation's surface water withdrawals. Public supply and industrial were the next largest users of surface water.

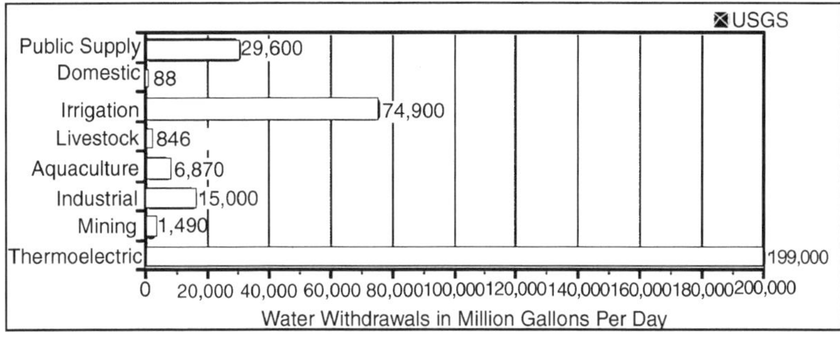

Fig. Total Surface-Water Withdrawals in the United States, 2005.

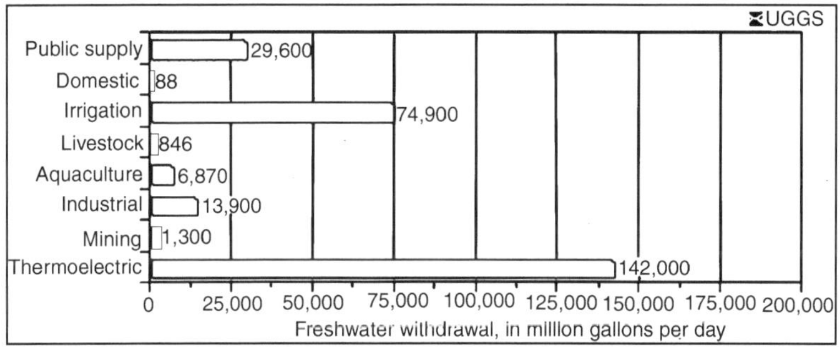

Fig. Total Surface Freshwater Withdrawals in the United States, 2005.

The pie charts below show the percentage of total water that was used in 2005 for various water-use categories, broken out by surface water and groundwater.

For most categories, surface water is used more than groundwater, although this pattern varies geographically across the United States. Domestic water use is almost exclusively groundwater, whereas the water used to produce electricity comes totally from surface water.

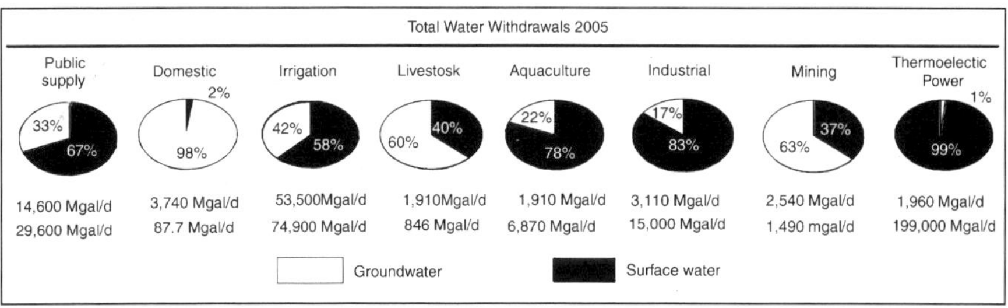

AQUEDUCTS MOVE WATER

If you live in an area where ample rain falls all year, you won't see many aqueducts like the ones pictured here. But there are many areas of the world, such as the western United States, where much less rainfall occurs and it may only occur during certain times of the year. Large cities and communities in the dry areas need lots of water, and nature doesn't always supply it to them.

Some parts of the western U.S. do have ample water supplies, though. So, some states have developed ways of moving water from the place of ample supply to the thirsty areas. Engineers have built aqueducts, or canals, to move water, sometimes many hundreds of miles. Actually, aqueducts aren't a high-tech modern invention—the ancient Romans had aqueducts to bring water from the mountains above Rome, Italy to the city.

Can you see something about the aqueduct in these depictions that causes some water to be lost in transit? In places where the climate is hot and dry, a certain portion of the water flowing in the aqueduct is bound to evaporate. It would be more efficient to cover the aqueduct to stop loss by evaporation, but the cost of covering it must be weighed against the value of the evaporated water.

RAIN

Rain and snow are key elements in the Earth's water cycle, which is vital to all life on Earth. Rainfall is the main way that the water in the skies comes down to Earth, where it fills our lakes and rivers, recharges the underground aquifers, and provides drinks to plants and animals. Fortunately for everyone, water is a renewable resource that moves in a cycle with neither beginning nor end. Water vapour condenses and returns to Earth as precipitation, once again replenishing reservoirs, lakes, rivers, and other sources of water and providing the moisture required by plants and animals's

The amount of precipitation that falls around the world may range from less than 0.1 inch per year in some deserts to more than 900 inches per year in the tropics. One of the driest spots on Earth is Lquique, Chile, where no rain fell for a period of 14 years. The world's wettest spot, as shown by data collected from a rainfall gage operated by the U.S. Geological Survey, is on Mt. Waialeale, Hawaii, where an average of more than 451 inches of rain falls each year, and where more than 642 inches fell from July 1947 to July 1948. Although Mt. Waialeale averages slightly more rain per year, Cherrapunji, India, holds the single year record of 905 inches measured in 1861.

By contrast, the conterminous United States receives enough precipitation during an average year to cover the States to a depth of about 30 inches. This is equivalent to about 1,430 cubic miles of water each year and would weigh about 6.6 billion tons. What happens to the water after it reaches the ground depends upon many factors such as the rate of rainfall, topography, soil condition, density of vegetation, temperature, and the extent of urbanization.

For example, the direct run-off in a highly urbanized area is relatively great, not only because of the density of roofs and impermeable pavements permits less rain to infiltrate the ground, but also because storm-sewer systems carry more water directly to the streams and lakes. In a more natural or undeveloped area, the direct run-off would be considerable less. In the United States, an average of some 70 per cent of the annual precipitation returns to the atmosphere by evaporation from land and water surfaces and by transpiration from vegetation. The remaining 30 per cent eventually reaches a stream, lake, or ocean, partly by overland run-off during and immediately after rain, and partly by a much slower route through the natural ground–water reservoir.

Much of the rain that enters the ground filters down into subsurface water-bearing rocks and eventually reaches lakes, streams, and rivers where these surface-water bodies intercept the aquifers.

The portion of the precipitation that reaches the streams produces an average annual streamflow in the United States of approximately 1,200 billion gallons a day. By comparison, the Nations's homes, farms, and factories withdraw and use about 400 billion gallons a day.

How Much Wateralls During a Rainstorm?

Have you ever wondered how much water falls onto your yard during a rainstorm? Using a 1-inch rainstorm as an example, the table below gives example of how much water falls during your storm for various land areas.

Table. Amount of Water Received When an inch of Rain Occurs

Area	Area (Square miles)	Area (Square kilometers)	Amount of water (gallons)	Amount of water (litres)
1 acre	.00156	.004	27,154 gallons	102,789 litres
1 square mile	1	2.6	17.38 million gallons	65.78 million litres
Washington,	61.4	159	1.07 billion gallons	4.04 billion litresDC
United States	3,537,438	9,161,922	61,474 billion gallons	232,700 billion litres

There are 640 acres in a square mile. Once on the land, rainfall either seeps into the ground or becomes run-off, which flows into rivers and lakes.

What happens to the rain after it falls depends on many factors such as:

- *The rate of rainfall*: A lot of rain in a short period tends to run off the land into streams rather than soak into the ground.
- *The topography of the land*: Topography is the lay of the land—the hills, valleys, mountains, and canyons. Water falling on unlevel land drains downhill until it becomes part of a stream, finds a hollow place to accumulate, like a lake, or soaks into the ground.
- *Soil conditions*: There is a lot of dense clay in the southeastern United States that rain has a hard time soaking into. Contrast that to the sandy soils in more desert areas, which allow water to quickly be absorbed, at least initially.
- *Density of vegetation*: It has long been known that plant growth helps decrease erosion caused by flowing water. If you look at hills without vegetation you'll see gullies dug out by running water. Land with

plant cover slows the speed of the water flowing on it and thus helps to keep soil from eroding.

- *Amount of urbanization*: As a city is being built, a lot of money and construction goes into moving water out of built-up areas. Roads, pavement, and parking lots create impervious areas where water can no longer seep into the ground. Rather, water is funneled into creeks and streams that were never meant by nature to handle so much run-off. This can cause problems in urban areas.

The table below gives example of how much water falls within the city limits of selected cities when one inch of rainfall occurs.

Table. Amount of Water Received When an inch of Rain Occurs

City	Area (Square miles)	Amount of water (Million gallons)
Atlanta, GA	131.7	2,289
Baltimore, MD	80.8	1,404
Chicago, IL	227.1	3,947
Cincinnati, OH	78.0	1,356
Denver, CO	153.4	2,666
Detroit, MI	138.8	2,412
Honolulu, HI	85.7	1,489
Houston, TX	579.4	10,069
Jacksonville, FL	757.7	13,168
Louisville, KY	62.1	1,079
Milwaukee, WI	96.1	1,670
New Orleans, LA	180.6	3,139
New York, NY	303.3	5,271
Philadelphia, PA	135.1	2,348
Salt Lake City, UT	109.1	1,906
Seattle, WA	83.9	1,458
Washington, DC	61.4	1,067

Consider for a moment how much rainwater some cities may receive during a year. for example, Atlanta, Ga. averages about 45 inches of precipitation per year; multiplying this by the 2.36 billion gallons shown in the table as the number of gallons in 1 inch reveals that some 106 billion gallons of water fall on Atlanta in an average year. In a city the size of Atlanta, the per capita water use is about 110 gallons per day or 40,150 gallons per year. Thus, the water from a year's precipitation, if it could be collected and stored without evaporation loss, would supply the needs of about 2,640,000.

Water Equivalents (Approximate)

The following equivalents show the relationship between the volume and weight of water and between the volume and speed of flowing water.

Volume and weight:

- One inch of rain falling on 1 acre of ground is equal to about 27,154 gallons and weighs about 113 tons.
- An inch of snow falling evenly on 1 acre of ground is equivalent to about 2,715 gallons of water. However, based upon the "rule-of-thumb" that 10 inches of snow is equal to 1 inch of water, can vary considerable, depending on whether the snow is heavy and wet, or powdery and dry. Heavy, wet snow has a very high water content—4 or 5 inches of this kind of snow contains about 1 inch of water. Thus, an inch of very wet snow over an acre might amount to more than 5,400 gallons of water, while an inch of powdery snow might yield only about 1,300 gallons.
- One acre–foot of water equals 326,000 gallons or 43,560 cubic feet of water, and weighs 2.7 million pounds.
- One cubic mile of water equals 1.1 trillion gallons, 147.2 billion cubic feet, or 3.38 million acre-feet, and weighs 9.2 trillion pounds.

Rate of flow (in a stream):

- *Water flowing at the steady rate of 1 gallon per minute is equivalent to*: 1,440 gallons per day; 0.00223 cubic foot per second; 192.7 cubic feet per day; or 0.00442 acre-foot of water per day.
- *Water flowing at the steady rate of 1 cubic foot per second is equivalent to*: 449 gallons per minute; 646,000 gallons per day; 86,400 cubic feet per day; or 1.98 acre-feet of water per day.
- *Water flowing at the steady rate of 1 acre-foot per day is equivalent to*: 226 gallons per minute; 326,000 gallons per day; 0.504 cubic foot per second; or 43, 560 cubic feet of water per day.
- *Water flowing at the steady rate of 1 cubic mile per day is equivalent to*: 764.6 million gallons per minute; 1.1 trillion gallons per day; 1.7 million cubic feet per day; or 3.38 million acre-feet of water per day.

FOLLOW A DROP THROUGH THE WATER CYCLE

You may be familiar with how water is always cycling around, through, and above the Earth, continually changing from liquid water to water vapour to ice. One way to envision the water cycle is to follow a drop of water around as it moves on its way. I could really begin this story anywhere along the cycle, but I think the ocean is the best place to start, since that is where most of Earth's water is. If the drop wanted to stay in the ocean then it shouldn't have been sunbathing on the surface of the sea. The heat from the sun found the drop, warmed it, and evaporated it into water vapour. It rose into the air and continued rising until strong winds aloft grabbed it and took it hundreds of miles until it was over land. There, warm updrafts coming from the heated land surface took the dropettes up even higher, where the air is quite cold. When the vapour got cold it changed back into it a liquid. If it was cold enough, it would have turned into tiny ice crystals, such as those that make up cirrus clouds. The vapour condenses on tiny particles of dust, smoke, and salt crystals to become part of a cloud.

After a while our drop combined with other drops to form a bigger drop and fell to the earth as precipitation. Earth's gravity helped to pull it down to the surface. Once it starts falling there are many places for water drops to go. Maybe it would land on a leaf in a tree, in which case it would probably evaporate and begin its process of heading for the clouds again. If it misses a leaf there are still plenty of places to go.

The drop could land on a patch of dry dirt in a flat field. In this case it might sink into the ground to begin its journey down into an underground aquifer as groundwater. The drop will continue moving as groundwater, but the journey might end up taking tens of thousands of years until it finds its way back out of the ground. Then again, the drop could be pumped out of the ground via a water well and be sprayed on crops. Or the well water containing the drop could end up in a baby's drinking bottle or be sent to wash a car or a dog. From these places, it is back again either into the air, down sewers into rivers and eventually into the ocean, or back into the ground.

But our drop may be a land–lover. Plenty of precipitation ends up staying on the earth's surface to become a component of surface water. If the drop lands in an urban area it might hit your house's roof, go down the gutter and your driveway to the curb. If a dog or squirrel doesn't lap it up it will run down the curb into a storm sewer and end up in a small creek. It is likely the creek will flow into a larger river and the drop will begin its journey back towards the ocean.

If no one interferes, the trip will be fast back to the ocean, or at least to a lake where evaporation could again take over. But, with billions of people worldwide needing water for most everything, there is a good chance that our drop will get picked up and used before it gets back to the sea. A lot of surface water is used for irrigation. Even more is used by power-production facilities to cool their electrical equipment. From there it might go into the cooling tower to be evaporated. Talk about a quick trip back into the atmosphere as water vapour—this is it. But maybe a town pumped the drop out of the river and into a water tank. From here the drop could go on to help wash your dishes, fight a fire, water the tomatoes, or flush your toilet. Maybe the local steel mill will grab the drop, or it might end up at a fancy restaurant mopping the floor. The possibilities are endless—but it doesn't matter to the drop, because eventually it will get back into the environment. From there it will again continue its cycle into and then out of the clouds, this time maybe to end up in the water glass of the President of the United States.

GROUNDWATER

Groundwater, pumped from beneath the earth's surface, is often cheaper, more convenient and less vulnerable to pollution than surface water. Therefore, it is commonly used for public water supplies. Groundwater provides the largest source of usable water storage in the United States. Underground reservoirs contain far more water than the capacity of all surface reservoirs and lakes, including the Great Lakes. In some areas, ground water may be the only option. Some municipalities survive solely on groundwater.

Hydrologists estimate the volume of water stored underground by measuring water levels in local wells and by examining geologic records from well-drilling to determine the extent, depth and thickness of water-bearing sediments and rocks. Before an investment is made in full-sized wells, hydrologists may supervise the drilling of test wells. They note the depths at which water is encountered and collect samples of soils, rock and water for laboratory analyses. They may run a variety of geophysical tests on the completed hole, keeping and accurate log of their observations and test results. Hydrologists determine the most efficient pumping rate by monitoring the extent that water levels drop in the pumped well and in its nearest neighbours. Pumping the well too fast could cause it to go dry or could interfere with neighbouring wells. Along the coast, overpumping can cause saltwater intrusion. By plotting and analysing these data, hydrologists can estimate the maximum and optimum yields of the well.

Polluted ground water is less visible, but more insidious and difficult to clean up, than pollution in rivers and lakes. Ground water pollution most often results from improper disposal of wastes on land. Major sources include industrial and household chemicals and garbage landfills, industrial waste lagoons, tailings and process wastewater from mines, oil field brine pits, leaking underground oil storage tanks and pipelines, sewage sludge and septic systems. Hydrologists provide guidance in the location of monitoring wells around waste disposal sites and sample them at regular intervals to determine if undesirable leachate—contaminated water containing toxic or hazardous chemicals—is reaching the ground water. In polluted areas, hydrologists may collect soil and water samples to identify the type and extent of contamination.

The chemical data then are plotted on a map to show the size and direction of waste movement. In complex situations, computer modeling of water flow and waste migration provides guidance for a clean-up programme. In extreme cases, remedial actions may require excavation of the polluted soil. Today, most people and industries realise that the amount of money invested in prevention is far less than that of cleanup. Hydrologists often are consulted for selection of proper sites for new waste disposal facilities.

The danger of pollution is minimized by locating wells in areas of deep ground water and impermeable soils. Other practices include lining the bottom of a landfill with watertight materials, collecting any leachate with drains, and keeping the landfill surface covered as much as possible. Careful monitoring is always necessary.

WATER CYCLE

Groundwater makes up about twenty per cent of the world's fresh water supply, which is about 0.61 per cent of the entire world's water, including oceans and permanent ice. Global groundwater storage is roughly equal to the total amount of freshwater stored in the snow and ice pack, including the north and south poles.

This makes it an important resource which can act as a natural storage that can buffer against shortages of surface water, as in during times of drought. Groundwater is naturally replenished by surface water from precipitation, streams, and rivers when this recharge reaches the water table. Groundwater can be a long-term 'reservoir' of the natural water cycle as opposed to short-term water reservoirs like the atmosphere and fresh surface water.

The Great Artesian Basin in central and eastern Australia is one of the largest confined aquifer systems in the world, extending for almost 2 million km^2. By analysing the trace elements in water sourced from deep underground, hydrogeologists have been able to determine that water extracted from these aquifers can be more than 1 million years old.

By comparing the age of groundwater obtained from different parts of the Great Artesian Basin, hydrogeologists have found it increases in age across the basin. Where water recharges the aquifers along the Eastern Divide, ages are young. As groundwater flows westward across the continent, it increases in age, with the oldest groundwater occurring in the western parts. This means that in order to have travelled almost 1000 km from the source of recharge in 1 million years, the groundwater flowing through the Great Artesian Basin travels at an average rate of about 1 metre per year.

ISSUES

Certain problems have beset the use of groundwater around the world. Just as river waters have been over–used and polluted in many parts of the world, so too have aquifers. The big difference is that aquifers are out of sight. The other major problem is that water management agencies, when calculating the 'sustainable yield' of aquifer and river water, have often counted the same water twice, once in the aquifer, and once in its connected river. This problem, although understood for centuries, has persisted, partly through inertia within government agencies. In Australia, for example, prior to the statutory reforms initiated by the Council of Australian Governments water reform framework in the 1990s, many Australian States managed groundwater and surface water through separate government agencies, an approach beset by rivalry and poor communication.

The time lags inherent in the dynamic response of groundwater to development have generally been ignored by water management agencies, decades after scientific understanding of the issue was consolidated. In brief, the effects of groundwater overdraft may take decades or centuries to manifest themselves. In a classic study in 1982, Bredehoeft and colleagues modelled a situation where groundwater extraction in an intermontane basin withdrew the entire annual recharge, leaving 'nothing' for the natural groundwater—dependent vegetation community. Even when the borefield was situated close to the vegetation, 30 per cent of the original vegetation demand could still be met by the lag inherent in the system after 100 years. By year 500 this had reduced to 0 per cent, signalling complete death of the groundwater-dependent vegetation.

The science has been available to make these calculations for decades; however water management agencies have generally ignored effects which will appear outside the rough timeframe of political elections. Marios Sophocleous argued strongly that management agencies must define and use appropriate timeframes in groundwater planning. This will mean calculating groundwater withdrawal permits based on predicted effects decades, sometimes centuries in the future.

As water moves through the landscape it collects soluble salts, mainly sodium chloride. Where such water enters the atmosphere through evapotranspiration, these salts are left behind.

In irrigation districts, poor drainage of soils and surface aquifers can result in water tables coming to the surface in low-lying areas. Major land degradation problems of soil salinity and waterlogging result, combined with increasing levels of salt in surface waters. As a consequence, major damage has occurred to local economies and environments.

First, flood mitigation schemes, intended to protect infrastructure built on floodplains, have had the unintended consequence of reducing aquifer recharge associated with natural flooding. Second, prolonged depletion of groundwater in extensive aquifers can result in land subsidence, with associated infrastructure damage–as well as saline intrusion. Fourth, draining acid sulphate soils, often found in low-lying coastal plains, can result in acidification and pollution of formerly freshwater and estuarine streams.

Another cause for concern is that groundwater drawdown from over-allocated aquifers has the potential to cause severe damage to both terrestrial and aquatic ecosystems–in some cases very conspicuously but in others quite imperceptibly because of the extended period over which the damage occurs.

Overdraft

Groundwater is a highly useful and often abundant resource. However, over-use, or overdraft, can cause major problems to human users and to the environment. The most evident problem is a lowering of the water table beyond the reach of existing wells. Wells must consequently be deepened to reach the groundwater; in some places the water table has dropped hundreds of feet because of extensive well pumping. In the Punjab region of India, for example, groundwater levels have dropped 10 meters since 1979, and the rate of depletion is accelerating. A lowered water table may, in turn, cause other problems such as groundwater–related subsidence and saltwater intrusion. Groundwater is also ecologically important. The importance of groundwater to ecosystems is often overlooked, even by freshwater biologists and ecologists. Groundwaters sustain rivers, wetlands and lakes, as well as subterranean ecosystems within karst or alluvial aquifers.

Not all ecosystems need groundwater, of course. Some terrestrial ecosystems—for example, those of the open deserts and similar arid environments—exist on irregular rainfall and the moisture it delivers to the soil, supplemented by moisture in the air. While there are other terrestrial ecosystems in more hospitable environments where groundwater plays no central role, groundwater is in fact fundamental to many of the world's major ecosystems.

Water flows between groundwaters and surface waters. Most rivers, lakes and wetlands are fed by, and feed groundwater, to varying degrees. Groundwater feeds soil moisture through percolation, and many terrestrial vegetation communities depend directly on either groundwater or the percolated soil moisture above the aquifer for at least part of each year. Hyporheic zones and riparian zones are examples of ecotones largely or totally dependent on groundwater.

Aquifer drawdown or overdrafting and the pumping of fossil water increases the total amount of water within the hydrosphere subject to transpiration and evaporation processes, thereby causing accretion in water vapour and cloud cover, the primary absorbers of infrared radiation in the Earth's atmosphere. Adding water to the system has a forcing effect on the whole earth system, an accurate estimate of which hydrogeological fact is yet to be quantified.

A Solution to Over-use of Groundwater

Irrigated agriculture is a big business in Asia. It accounts for 70 per cent of the world's irrigated land and 73 per cent of water used each year by the farming industry. As ageing large-scale surface irrigation schemes have become increasingly inefficient, and farmers have begun growing a wider range of crops requiring water on demand, the number of groundwater wells in India has exploded.

In 1960, there were fewer than 100,000 such wells; by 2006 the figure had risen to nearly 12 million. In India, a possible solution to over-use of groundwater is emerging, known as 'groundwater recharge'. It involves capturing rainwater that would otherwise run-off, and using it to refill aquifers.

Since 2000, the International Water Management Institute has been working with the Indian authorities to help improve the availability of water for agriculture in India. In 2006, India's finance minister invited IWMI to submit policy recommendations based on its research on groundwater depletion.

One of the key recommendations was to instigate a programme of recharging groundwater across the 65 per cent of India that has hard-rock aquifers.

As a result, the Indian government allocated ₹1800 crore to fund dug–well recharge projects in 100 districts within seven states where water stored in hard–rock aquifers has been over-exploited.

These geological formations have a much lower capacity to store rainwater than alluvial areas with porous sand or clay rocks, hence being given priority. The money is sufficient to fund seven million structures to be installed on dug–wells to divert monsoon run-off.

The structures include a de-siltation chamber, plus pipes to collect surplus rainwater and divert de-silted water from the chamber to the well. As of the end of November 2009, funds amounting to ₹216.98 crore had been released to the concerned states. Subsidies had been released to 566,637 beneficiaries.

Subsidence

Subsidence occurs when too much water is pumped out from underground, deflating the space below the above–surface, and thus causing the ground to actually collapse. The result can look like craters on plots of land.

This occurs because in its natural equilibrium state, the hydraulic pressure of groundwater in the pore spaces of the aquifer and the aquitard supports some of the weight of the overlying sediments.

When groundwater is removed from aquifers by excessive pumping, pore pressures in the aquifer drop and compression of the aquifer may occur. This compression may be partially recoverable if pressures rebound, but much of it is not.

When the aquifer gets compressed it may cause land subsidence, a drop in the ground surface. The city of New Orleans, Louisiana, is actually below sea level today, and its subsidence is partly caused by removal of groundwater from the various aquifer/aquitard systems beneath it.

In the first half of the 20th century, the city of San Jose, California, dropped 13 feet from land subsidence caused by overpumping; this subsidence has been halted with improved groundwater management.

Seawater Intrusion

Generally, in very humid or undeveloped regions, the shape of the water table mimics the slope of the surface. The recharge zone of an aquifer near the seacoast is likely to be inland, often at considerable distance. In these coastal areas, a lowered water table may induce sea water to reverse the flow towards the land. Sea water moving inland is called a saltwater intrusion. Alternatively, salt from mineral beds may leach into the groundwater of its own accord.

Mining

Sometimes the water movement from the recharge zone to the place where it is withdrawn may take centuries. When the usage of water is greater than the recharge, it is referred to as mining water. Under those circumstances it is not a renewable resource.

Pollution

Water pollution of groundwater, from pollutants released to the ground that can work their way down into groundwater, can create a contaminant plume within an aquifer. Movement of water and dispersion within the aquifer spreads the pollutant over a wider area, its advancing boundary often called a plume edge, which can then intersect with groundwater wells or daylight into surface water such as seeps and springs, making the water supplies unsafe for humans and wildlife. The interaction of groundwater contamination with surface waters is Analysed by use of hydrology transport models.

The stratigraphy of the area plays an important role in the transport of these pollutants. An area can have layers of sandy soil, fractured bedrock, clay, or hardpan. Areas of karst topography on limestone bedrock are sometimes vulnerable to surface pollution from groundwater. Earthquake faults can also be entry routes for downward contaminant entry. Water table conditions are of great importance for drinking water supplies, agricultural irrigation, waste disposal, wildlife habitat, and other ecological issues.

In the US, upon commercial real estate property transactions both groundwater and soil are the subjects of scrutiny, with a Phase I Environmental Site Assessment normally being prepared to investigate and disclose potential pollution issues. In the San Fernando Valley of California, Real estate contracts for property transfer below the Santa Susana Field Laboratory and eastward have clauses releasing the seller from liability for groundwater contamination consequences from existing or future water pollution of the Valley Aquifer.

Love Canal was one of the most widely known examples of groundwater pollution. In 1978, residents of the Love Canal neighbourhood in upstate New York noticed high rates of cancer and an alarming number of birth defects. This was eventually traced to organic solvents and dioxins from an industrial landfill that the neighbourhood had been built over and around, which had then infiltrated into the water supply and evaporated in basements to further contaminate the air. Eight hundred families were reimbursed for their homes and moved, after extensive legal battles and media coverage.

Another example of widespread groundwater pollution is in the Ganges Plain of northern India where severe contamination of groundwater by naturally occurring arsenic affects 25 per cent of water wells in the shallower of two regional aquifers. The pollution occurs because aquifer sediments contain organic matter that generates anaerobic conditions in the aquifer.

These conditions result in the microbial dissolution of iron oxides in the sediment and thus the release of the arsenic, normally strongly bound to iron oxides, into the water. As a consequence, arsenic-rich groundwater is often iron-rich, although secondary processes often obscure the association of dissolved arsenic and dissolved iron.

GROUNDWATER QUALITY

Just because you have a well that yields plenty of water doesn't mean you can go ahead and just take a drink. Because water is such an excellent solvent it can contain lots of dissolved chemicals. And since groundwater moves through rocks and subsurface soil, it has a lot of opportunity to dissolve substances as it moves. For that reason, groundwater will often have more dissolved substances than surface water will.

Even though the ground is an excellent mechanism for filtering out particulate matter, such as leaves, soil, and bugs, dissolved chemicals and gases can still occur in large enough concentrations in groundwater to cause problems. Undergroundwater can get contaminated from industrial, domestic, and agricultural chemicals from the surface. This includes chemicals such as pesticides and herbicides that many homeowners apply to their lawns.

Contamination of groundwater by road salt is of major concern in northern areas of the United States. Salt is spread on roads to melt ice, and, with salt being so soluble in water, excess sodium and chloride is easily transported into the subsurface groundwater.

The most common water–quality problem in rural water supplies is bacterial contamination from septic tanks, which are often used in rural areas that don't have a sewage-treatment system. Effluent from a septic tank can percolate down to the water table and maybe into a homeowner's own well. Just as with urban water supplies, chlorination may be necessary to kill the dangerous bacteria. The U.S. Geological Survey is involved in monitoring the Nation's ground-water supplies. A national network of observation wells exists to measure regularly the water levels in wells and to investigate water quality.

GROUNDWATER FORMATION

Groundwater is fresh water in the rock and soil layers beneath Earth's land surface. Some of the precipitation (rain, snow, sleet, and hail) that falls on the land soaks into Earth's surface and becomes groundwater. Water-bearing rock layers called aquifers are saturated (soaked) with groundwater that moves, often very slowly, through small openings and spaces.

This groundwater then returns to lakes, streams, and marshes (wet, low-lying land with grassy plants) on the land surface via springs and seeps (small springs or pools where groundwater slowly oozes to the surface). Groundwater makes up more than one-fifth (22 per cent) of Earth's total fresh water supply, and it plays a number of critical hydrological (water-related), geological and biological roles on the continents.

Soil and rock layers in groundwater recharge zones (a entry point where water enters an aquifer) reduce flooding by absorbing excess run-off after heavy rains and spring snowmelts.

Aquifers store water through dry seasons and dry weather, and groundwater flow carries water beneath arid (dry) deserts and semi-arid grasslands. Groundwater discharge replenishes streams, lakes, and wetlands on the land surface and is especially important in arid regions that receive limited rainfall. Flowing groundwater interacts with rocks and minerals in aquifers, and carries dissolved rock-building chemicals and biological nutrients. Vibrant communities of plants and animals (ecosystems) live in and around groundwater springs and seeps.

Almost all of the fresh liquid water that is readily available for human use comes from underground. (The bulk of Earth's fresh water is frozen in ice in the North and South Pole regions. Water in streams, rivers, lakes, wetlands, the atmosphere, and within living organisms makes up only a tiny portion of Earth's fresh water.) For thousands of years, humans have used groundwater from springs and shallow wells to fill drinking water reservoirs, and water livestock and crops.

Today, human water needs far exceed surface water supplies in many regions, and Earth's rapidly-growing human population relies heavily upon groundwater to meet its ever larger demand for clean, fresh water.

AQUIFERS: FRESH WATER UNDERGROUND

An aquifer is a body of rock or soil that yields water for human use. Most aquifers are water-saturated layers of rock or loose sediment.

With the exception of a few aquifers that have water-filled caves within them, aquifers are not underground lakes or holding tanks, but rather rock "sponges" that hold groundwater in tiny cracks, cavities, and pores (tiny openings in which a liquid can pass) between mineral grains (rocks are made of minerals).

The total amount of empty pore space in the rock material, called its porosity, determines the amount of groundwater the aquifer can hold. Materials like sand and gravel have high porosity, meaning that they can absorb a high amount of water.

Rocks like granite, marble, and limestone have low porosity, and make poor groundwater reservoirs. Aquifers must have high permeability in addition to high porosity. Permeability is the ability of the rock or other material to allow water to pass through it. The pore space in permeable materials is interconnected throughout the rock or sediment, allowing groundwater to move freely through it.

Some high–porosity materials, like mud and clay, have very low permeability. They soak up and hold water, but don't release it easily to wells or other groundwater discharge points, so they are not good aquifer materials. Sandstone, limestone, fractured granite, glacial sediment, loose sand, and gravel are examples of materials that make good aquifers. Water enters aquifers by seeping into the land surface at entry points called recharge zones and leaves at exit points called discharge zones. (Some aquifers discharge into the ocean.)

Influent or "water-losing" streams, ponds, or lakes are bodies of surface water in recharge zones that contribute groundwater from their water supply. Groundwater flows into effluent or "watergaining" streams and ponds in discharge zones. For the water level in an aquifer to remain constant, the amount of water entering at recharge zones must equal the amount leaving at discharge zones. (Imagine a bucket punched with holes under a dripping faucet. If water drips in at the same rate that it drips out, the water level stays the same.)

If water discharges or is pumped from an aquifer more quickly than it recharges, the groundwater level will fall. The time an average water molecule spends within an aquifer is called its residence time. Water in some fast–flowing aquifers spends only a few days underground, while other rock layers can hold water for ten thousand years. Average aquifers have residence times of about two hundred years.

The Water Table and Unconfined Aquifers

Water enters aquifers by moving slowly down through a layer of surface rocks and soil whose pore spaces are partially filled with air (zone of infiltration). The water continues moving downwards until it reaches a level where all the pore spaces are completely filled with water (zone of saturation).

The top of the zone of saturation is called the water table. In some wet, lowland regions, southern Florida for example, the water may be only a few feet (meters) below the surface. In others, like the American Southwest, water–saturated rocks may be hundreds of feet below the land surface. Groundwater reservoirs that have uniform rock or soil properties (porosity and permeability) throughout are called unconfined aquifers. The water table forms the upper surface of an unconfined aquifer. The shape of the water table in an unconfined aquifer mirrors the shape of the land surface, but its slopes are gentler.

In temperate (moderate) climates that receive moderate amounts of groundwater–replenishing rainfall, water infiltrates into unconfined aquifers in hilltop recharge zones and discharges into effluent streams and ponds in low areas where the water table intersects the land surface. Water will only rise to the level of the water table in a well, so a pump or bucket is required to extract water from an unconfined aquifer.

Confined Aquifers and Artesian Flow

Confined aquifers are pressurized groundwater reservoirs that lie beneath layers of non-permeable rock (granite, shale) or sediment (clay). Groundwater enters a confined aquifer in recharge zones beyond the uphill edges of the confining layer and discharges beyond the downhill edges. Groundwater trapped beneath an impermeable barrier cannot rise to the height of the water table, so pressure builds up in confined aquifers. Artesian wells are wells drilled in confined aquifers where the pressure is great enough to make water flow at the surface.

Lakes are large inland bodies of fresh or saline (salty) water. Lakes form in places where water collects in low areas or behind natural or man-made dams (barriers constructed to contain the flow of water). Some lakes are fed by streams (natural bodies of flowing freshwater), and some form where groundwater (water flowing in rock layers beneath the land surface) discharges onto the land surface. Water leaves lakes by flowing into outlet streams, infiltrating (soaking in) into groundwater reservoirs called aquifers, and by evaporating into the atmosphere (mass of air surrounding Earth). Lakes vary in size from large lakes such as the Great Lakes of North America, to small mountain lakes. Lakes are larger than ponds, which are small bodies of fresh water that are shallow enough for rooted plants to grow.

The study of ecology (relationships between living organisms and their environment) in lakes, inland seas, and wetlands is called limnology. Lakes store only a tiny percentage of Earth's fresh water. They are, however, an extremely important water resource for humans.

Freshwater lakes provide water for agricultural irrigation (watering), industrial processes, municipal uses, and residential water supplies. People who live in the continental interiors use lakes for fishing and recreation, and very large lakes have important shipping and transportation routes.

Humans also construct artificial lakes, called reservoirs, by building dams across rivers. In addition to providing the benefits of natural lakes, reservoirs also store water for specific communities, control floods, and generate hydroelectricity (electricity generated from water power).

HOW LAKES FORM AND DISAPPEAR

Earth scientists who study water on the continents (hydrologists and hydrogeologists) see lakes as temporary reservoirs within stream and groundwater systems. All water that falls as precipitation (rain, snow, sleet, hail) on the land surface of continents eventually makes its way to the ocean or evaporates back into the atmosphere. Water collects in lakes because it enters more rapidly than it escapes, but it is never permanently trapped there. As in a tub with a running faucet and an open drain, individual water molecules (smallest unit of water, each containing two hydrogen atoms and one oxygen atom) are constantly entering and escaping. After arriving in a lake, an average water molecule spends about one hundred years before moving to a new reservoir. (The time that an average water molecule spends in a reservoir is called it residence time. Water resides for about two weeks in rivers, forty years in glaciers (slow moving mass of ice), and between two hundred and ten thousand years in groundwater reservoirs.)

To geologists (Earth scientists), lakes are temporary features. Stream-fed lakes within stream systems are destined for destruction. Every stream seeks to create a constant slope, called a graded profile, between where the stream's waters begin and ends by eroding (wearing away) and depositing sediment (particles of sand, silt, and clay) along its course.

When a natural or man-made obstruction blocks a stream, such as a river, streams deposit sediment in the lake or reservoir behind the obstruction and erode away in front of it. Eventually, the dam will collapse, and the lake will empty. Lakes that fill depressions and have no outlets fill when the regional climate becomes wetter or when warm periods melt mountain snows and glacial ice. They evaporate away during periods of dry weather and dryer climate. It may take thousands, or even tens of thousands of years, but lakes eventually drain, collapse, or dry up.

LAKE LAYERS AND OVERTURNS

Contrary to their common image as evenly mixed pools of unmoving water, lakes are complex, dynamic bodies of moving surface water. Lake water varies within the lake; its temperature, chemical content, light infiltration, and biological habitats vary from top to bottom and side to side.

Furthermore, the vertical layering (stratification), horizontal variations, and circulation patterns within lakes change over time. Waves, currents (a moving mass of water), and even tides affect circulation of water within lakes. Lakes are thermally stratified (layered just as to temperature); they have layers of warm and cool water that are separated by layers where the temperature changes (thermoclines). Like the oceans, many lakes have a thin layer of warm surface water, and a thicker layer of cool deep water that is separated by a thermocline layer. Wind generates currents on lake surfaces and creates some mixing. Unlike the oceans, however, many lakes have seasonal overturns that mix their waters. water is denser than solid water (ice). Water reaches its maximum density at 39°F (4°C). Because of this odd property, the warm less-dense water rises, the cool denser water sinks, ice floats, and lakes overturn.

Lakes that are ice-covered for part of the year undergo overturns that partially or completely mix their waters. Many lakes in temperate (moderate temperatures) regions like the northern United States overturn and mix completely twice a year (dimictic lakes).

During the warm summer months, these lakes have a usual temperature profile with warm surface waters, a thermocline, and cool bottom water. In the fall, when the surface water cools down to 39°F (4°C), it becomes denser than the water underneath it and the surface layer sinks to the bottom.

The bottom water rises to the surface, and the lake overturns. Over the winter, the bottom water is the warmest, and the frozen surface water is the coldest. (Plants and animals survive the winter on the lake floor in the chilly, but not frozen, bottom water.) In the spring, when the ice melts, and the water warms to 39°F (4°C), it sinks, and the lake overturns again. Limnologists classify lakes just as to the number of mixing events they undergo each year.

Lake type classifications have the root term mictic, meaning "to mix," and include:

- *Oligomictic*: Warm, ice-free lakes that rarely mix. They are warmest at the top, and coolest at the bottom. Tropical, oligomictic lakes have warm bottom water and very warm surface water. They rarely overturn because their water does not near 39°F (4°C).
- *Meromictic*: Warm, ice–free lakes that mix incompletely. These are deep lakes that are warmest at the top, and coolest at the bottom.
- *Monomictic/dimictic/polymictic*: Lakes with seasonal ice covers that overturn and mix completely once (monomictic), twice (dimictic), or many times per year (polymictic). Lake overturns are the norm in temperate regions, but local conditions affect the timing and number of overturns in specific lakes.
- *Amictic*: Lakes that never overturn because they are icecovered throughout the year. These lakes exist near the North and South Poles and atop very high mountains. They have cold bottom water that hovers near 39°F (4°C) and frozen surface water.

LAKE CHEMISTRY: SALINE LAKES

Many of Earth's largest and most important lakes contain salt water. All surface water contains some dissolved chemicals, called salts. Groundwater, streams, and freshwater lakes all contain the chemical components of rocks and minerals. Humans can drink fresh water because our bodies can use or at least tolerate the types and concentrations of dissolved chemicals it contains.

Salt water, on the other hand, has a very high concentration of dissolved salts, and is undrinkable.

The Dead Sea, on the border between Israel and Jordan, is Earth's saltiest body of water. It is truly a dead sea because it is too salty to support life. Saline lakes generally form in arid (dry) regions where surface water evaporates quickly.

When water evaporates, the salts stay behind. Over time, the lake water becomes saltier. Some saline lakes, such as the Great Salt Lake, are all that remains of a much larger fresh water lake that has evaporated over time. Others, like the Caspian Sea in central Asia, began as saltwaterfilled ocean basins that have since closed. Saline lakes are often temporary features that fill during periods of wetter climate and then dry up when stream flow or groundwater discharge slows. Playa lakes are flat desert basins that occasionally fill with water.

Desert oases (watering holes) form and disappear with such regularity that thirsty travellers think they imagined them. The Great Salt Lake, Caspian Sea, Aral Sea, and Dead Sea are all presently evaporating. Over time, the dissolved chemicals become so concentrated in drying lakes that they bond together and form solid salt crystals. Thick layers of salt cover dry lake beds.

LAKE BIOLOGY

Lakes support rich communities of plants and animals (ecosystems) that have adapted to live within ever-changing conditions on lake beds, within the water column (water running from the surface to the lake floor, often showing differences in temperature, nutrients, etc.), and along lake shores. Lakes, like islands, are often closed systems that only rarely gain new species or individuals from other lakes.

Many lakes host groups of rare species that have evolved (changed over time) together in their specific lake. These ecosystems are rich and unique, but fragile. They have little defence against foreign predators or diseases. Human alterations and water pollution have threatened many lake species. Environmental groups and government agencies are presently attempting to protect and revive threatened lake species such as cichlids (rare doublejawed fish) that inhabit the lakes of the Great Rift Valley in east Africa. Lake organisms live in zones that are determined by the physical structure of their lakes such as the amount of available light, water depth, and distribution of nutrients. Most lake plants and animals live in shallow, well-lit surface waters called the euphotic zone.

Most plants depend on the Sun's energy to produce food by the chemical process of photosynthesis, and they cannot grow in water that is too deep or too cloudy for light to penetrate. Lake animals such as fish need oxygen that plants give off during photosynthesis, so they live mostly in the euphotic zone as well. Plants with roots grow in shallow water along edges of lakes where light reaches the lake floor (littoral zone) and floating plants perform photosynthesis in the open surface waters (limnetic zone). Oxygen-consuming bacteria inhabit the deepest, darkest parts of lakes (benthic zone) where dead plant and animal materials accumulate. Limnologists also classify lakes by the balance of organisms and nutrients in their waters.

Types include lakes that are described as oligotrophic, eutrophic, and mesotrophic:

- *Oligotrophic*: Nutrient poor lakes that support very few plants and animals. Oligotrophic lakes are typically cool, deep, and have very clear water. Very little organic (relating to or from living organisms) mud accumulates in oligotropic lakes, and they often have sand and gravel beds.
- *Eutrophic*: Lakes rich in plant nutrients that support abundant plant life in their surface waters. Their water is often clouded by microscopic plants, and their beds covered with thick layers of decaying plant material. Bacteria that live on the organic mud use up oxygen, and eutrophic lakes often have oxygen–poor deep water. Plants and bacteria eventually take over eutrophic lakes. They become oxygen-poor bogs and marshes where fish cannot live. Some chemicals that humans use, including fertilizers and detergents, cause a process called eutrophication when they run off into lakes, which causes the population of plants to increase to such an extent that eventually oxygen-starved fish die.
- *Mesotrophic*: Lakes with moderate amounts of nutrients and healthy, balanced communities of plants, animals and bacteria. Mesotrophic lakes receive adequate amounts of fresh water and nutrients, and seasonal overturns allow nutrient–poor and nutrient–rich layers to mix. Mesotrophic lakes are intermediate between crystal-clear, lifeless oligothrophic lakes and cloudy, muddy eutrophic lakes.

WHERE LAKES FORM: LAKE BASINS

Lakes form where water collects in depressions, or basins. Many lakes fill low areas created by plate tectonic movements (tectonic basins) and volcanic activity. (Plate tectonics is the movement of large, rigid pieces of Earth's outer rock shell called the lithosphere.) Retreating glaciers and ice sheets leave behind large basins and small depressions that fill with meltwater. Though flowing streams and rivers generally act to fill in and drain lake basins, other sedimentary processes can create landscape depressions and natural dams that confine water in lakes.

Lakes in Tectonic Basins

Rift valley lakes fill long, linear valleys within rift zones. (Rifts are areas where the continental lithosphere is stretching and beginning to break into pieces. They are the precursors of ocean basins.) A chain of large lakes including Tanganyika, Naivete, and Malawi follows the Great Rift Valley through eastern

Africa. Lake Victoria, the world's second-largest lake, lies between two branches of the rift valley. The Red Sea, Sea of Galilee, Dead Sea, and Gulf of Ababa fill the northern branches of the rift where it crosses the Arabian Peninsula in the Middle East. Russia's Lake Baikal, the world's deepest lake, fills an ancient, inactive rift valley in central Asia. Lakes also form in places where continents are moving towards each other.

The Black Sea, Caspian Sea, and Mediterranean Seas fill a closing ocean basin between Africa and Europe. When continents collide, water fills depressions in the landscape over folded and broken (faulted) rock layers that were caught between the land masses. Slopes that are too steep collapse and block rivers with natural dams. Blocks of uplifted, erosion-resistant rock form bedrock that holds back mountain lakes.

Volcanic Lakes

Volcanoes are mountains that form from eruptions of molten rock (lava) on the land surface. When a volcanic peak collapses into its emptied magma chamber (a pool or room of magna held under tremendous pressure within a volcano prior to a volcanic eruption), it forms a large circular basin called a caldera. (Craters, the small basins near the top of active volcanoes, sometimes also contain small lakes, but most significant volcanic lakes, including inaccurately-named Crater Lake, fill calderas.) Yellowstone Lake in Wyoming and Crater Lake in Oregon are examples of caldera lakes. Volcanic ash, mud, and lava flows also create natural dams in river valleys. A dam of volcanic rock confines Lake Tahoe in a high valley of the Sierra Mountains on the California–Nevada border.

Glacial Lakes

The thick continental ice sheets that covered northern North America, Europe, and Asia during the Pleistocene ice ages (a division of geologic time that lasted from two million to ten thousand years ago) left behind thousands of lake and ponds when they retreated about twenty thousand years ago.

The weight of the ice sheets pushed down on the continents, leaving broad basins that filled with melt water when they retreated. The Great Lakes of North America (Superior, Huron, Michigan, Erie, and Ontario) formed this way. Hundreds of lakes, such as the Winnipeg, Athabasca, Great Slave, and Great Bear cover the central and eastern provinces and territories of Canada that are still rebounding from their heavy ice load. Advancing glaciers also pile tall ridges of sediment, called moraines, at their toes (the end of extensions of glaciers along the ground). When glaciers retreat, moraines hold back meltwater. Small lakes and ponds also form in glacial depressions called kettles that form when blocks of ice buried in glacial sediment melt. Melting mountain glaciers feed many mountain lakes and glacial sediment traps streams and meltwater.

Groundwater Discharge Lakes

Water moving through pore spaces in rock and soil layers discharges on the land surface in places where the water table (level below which pore spaces are saturated with water) intersects the land surface. In regions with wet climates, the water table is near the land surface and ground water discharges in low spots. Groundwater chemically erodes limestone and other rocks and creates caves, cavities, sink holes, and collapse basins called karst features. Florida's many lakes, including Lake Okeechobee, are groundwater filled karst features.

HYDROPOWER AND THE ENVIRONMENT

HYDROPOWER IS NON-POLLUTING, BUT DOES HAVE ENVIRONMENTAL IMPACTS

Hydropower does not pollute the water or the air. However, hydropower facilities can have large environmental impacts by changing the environment and affecting land use, homes, and natural habitats in the dam area. Most hydroelectric power plants have a dam and a reservoir.

These structures may obstruct fish migration and affect their populations. Operating a hydroelectric power plant may also change the water temperature and the river's flow. These changes may harm native plants and animals in the river and on land.

Reservoirs may cover people's homes, important natural areas, agricultural land, and archeological sites. So building dams can require relocating people. Methane, a strong greenhouse gas, may also form in some reservoirs and be emitted to the atmosphere. (EPA Energy Kids)

RESERVOIR CONSTRUCTION IS "DRYING UP" IN THE UNITED STATES

Gosh, hydroelectric power sounds great—so why don't we use it to produce all of our power? Mainly because you need lots of water and a lot of land where you can build a dam and reservoir, which all takes a LOT of money, time, and construction.

In fact, most of the good spots to locate hydro plants have already been taken. In the early part of the century hydroelectric plants supplied a bit less than one-half of the nation's power, but the number is down to about 10 per cent today.

The trend for the future will probably be to build small-scale hydro plants that can generate electricity for a single community. As this chart shows, the construction of surface reservoirs has slowed considerably in recent years. In the middle of the 20th Century, when urbanization was occuring at a rapid rate, many reservoirs were constructed to serve peoples' rising demand for water and power. Since about 1980, the rate of reservoir construction has slowed considerably.

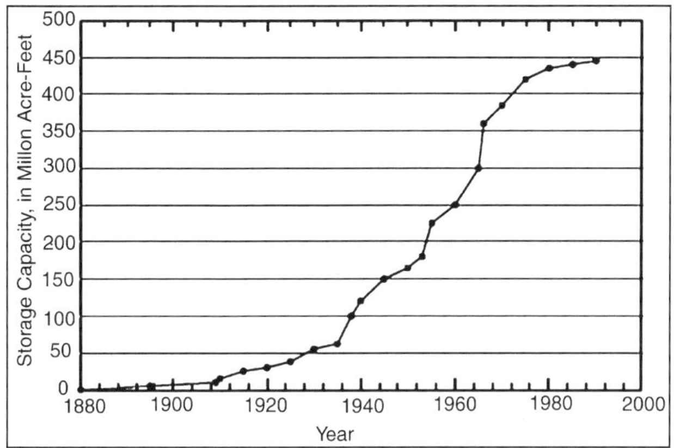

TYPICAL HYDROELECTRIC POWER PLANT

Hydroelectric energy is produced by the force of falling water. The capacity to produce this energy is dependent on both the available flow and the height from which it falls. Building up behind a high dam, water accumulates potential energy.

This is transformed into mechanical energy when the water rushes down the sluice and strikes the rotary blades of turbine. The turbine's rotation spins electromagnets which generate current in stationary coils of wire. Finally, the current is put through a transformer where the voltage is increased for long distance transmission over power lines.

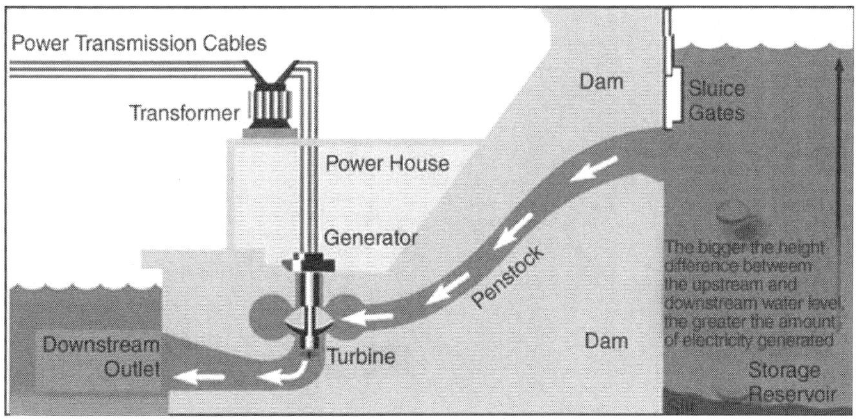

Fig. Hydroelectric Power Generation.

HYDROELECTRIC-POWER PRODUCTION IN THE UNITED STATES AND THE WORLD

In the United States, most states make some use of hydroelectric power, although, as you can expect, states with low topographical relief, such as Florida and Kansas, produce very little hydroelectric power.

But some states, such as Idaho, Washington, and Oregon use hydroelectricity as their main power source. in 1995, all of Idaho's power came from hydroelectric plants.

The hydroelectric power generation in 2006 for the leading hydroelectric-generating countries in the world. China has developed large hydroelectric facilities in the last decade and now lead the world in hydroelectricity usage. But, from north to south and from east to west, countries all over the world make use of hydroelectricity—the main ingredients are a large river and a drop in elevation.

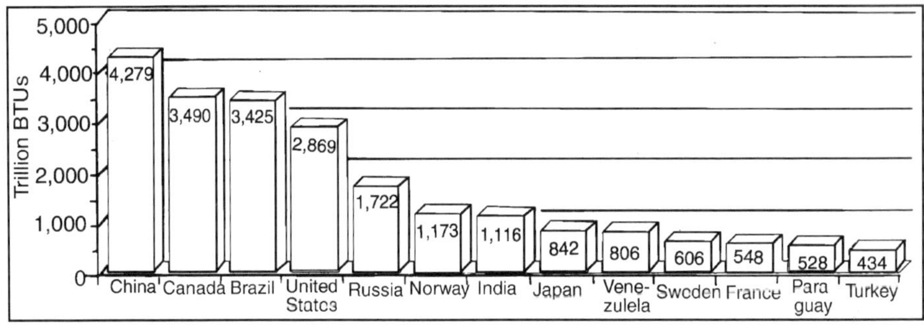

Fig. World Net Hydroelectric Power Generation (Trillion Btu) 2006.

7

Hydrology, Soil Water and Temperature

BRINGING ALL THE PIECES TOGETHER

The hydrologic cycle begins with the evaporation of water from the surface of the ocean. As moist air is lifted, it cools and water vapor condenses to form clouds. Moisture is transported around the globe until it returns to the surface as precipitation.

Once the water reaches the ground, one of two processes may occur; 1) some of the water may evaporate back into the atmosphere or 2) the water may penetrate the surface and become groundwater. Groundwater either seeps its way to into the oceans, rivers, and streams, or is released back into the atmosphere through transpiration. The balance of water that remains on the earth's surface is run-off, which empties into lakes, rivers and streams and is carried back to the oceans, where the cycle begins again.

Lake effect snowfall is good example of the hydrologic cycle at work. Below is a vertical cross-section summarizing the processes of the hydrologic cycle that contribute to the production of lake effect snow. The cycle begins as cold winds (horizontal blue arrows) blow across a large lake, a phenomena that occurs frequently in the late fall and winter months around the Great Lakes.

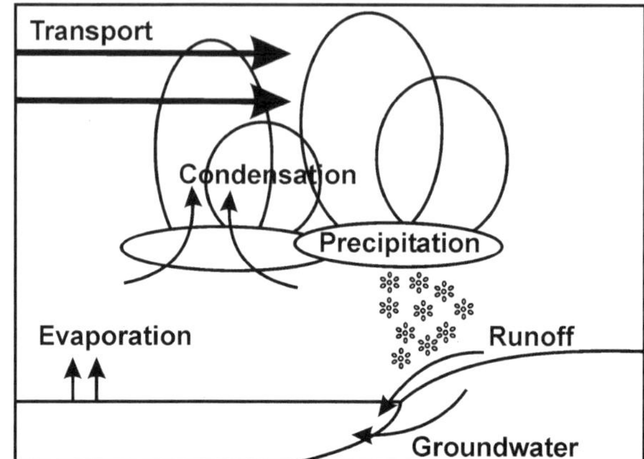

Evaporation of warm surface water increases the amount of moisture in the colder, drier air flowing immediately above the lake surface. With continued evaporation, water vapor in the cold air condenses to form ice-crystal clouds, which are transported towards shore.

By the time these clouds reach the shoreline, they are filled with snowflakes too large to remain suspended in the air and consequently, they fall along the shoreline as precipitation.

The intensity of lake effect snowfall can be enhanced by additional lifting due to the topographical features (hills) along the shoreline. Once the snow begins to melt, the water is either absorbed by the ground and becomes groundwater, or goes returns back to the lake as run-off.

SOIL WATER PROPERTIES

Two soil water properties are of the utmost importance for understanding, describing and managing soil water.

These properties are soil water content and soil water potential. Water potential was described in detail in the preceding lecture. Water content simply refers to how much water is present. For each soil, the relation between content and potential is the basis for that soil's ability to provide water to plants.

Water content describes how much water is in soil. This content can be expressed on the basis of soil and water mass or on soil and water volume. The mass method is also called the gravimetric method. Mass or gravimetric water content is designated by the symbol qm. Volumetric water content is designated by the symbol qv.

Mass water content is a ratio of water mass to dry soil mass. Normally it is best visualized as grams of water per gram of dry soil. Mass water content is best expressed as a fraction (and often incorrectly expressed as a per cent). The following example shows how to calculate qm.

Example: A moist soil weighs 120 g when dried the soil weighs 100 g the water mass was 20 g therefore, qm = 20 g water/100 g soil = 0.20 Rarely, as in an organic soil, qm can be greater than 1.

Volumetric water content is the volume of water per volume of soil. Volumetric water content can be expressed as fraction, but can correctly be expressed as a per cent. As the example below illustrates, qv can be calculated from qm by multiplying qm by bulk density and by the density of water.

Example:

$$\frac{Mass\ of\ water}{Mass\ of\ soil} = \frac{Mass\ of\ soil}{Mass\ of\ water} = \frac{voulume\ of\ water}{mass\ of\ water} = qv$$

$$r_B\ 1\ cc/1\ g$$

Volumetric water content is generally more useful than qm.

Depth of water in the field is another useful quantity. Remember that:

$$qv = \frac{Volume\ of\ water}{Volume\ of\ soil}$$

Note that volume of water = area x depth of water

and, volume of soil = area x depth of soil

Because the area is the same quantity for water and soil it cancels; therefore

$$qv = \frac{Depth\ of\ water}{Depth\ of\ soil}$$

Or, one can rewrite the equation such as:

$$\text{Depth of soil} \times qv = \text{Depth of water}$$

From this equation one can solve problems such as in the following example.

Example: how much water is in 50 cm of soil when qv = 0.15

$$50\ cm \times .15 = 7.5\ cm$$

Characteristic curves relate y to q. Each soil has a unique characteristic curve. From the relation described by the curve, one can determine:

- Water content at the permanent wilting point (PWP)
- Water content at field capacity (FC)
- Plant available water (PAW)

Plant available water is water that the soil will relinquish to the plant. Each soil has a particular qv at its FC. Field capacity is the amount of water a soil can hold against the pull of gravity.

As a working definition, it can be considered as the amount of water in a soil about 24 hours after a soaking rain. It occurs at a potential of -0.33 bars for clays and loams and -0.10 bars for sand. Typical qv values for soils at field capacity are:

- Sandy ~ 0.15
- Loamy ~ 0.30
- Clay ~ 0.40

Likewise, each soil has a particular qv at the PWP. A large array of plants can pull no harder on soil water than -15 bars. For this reason, -15 bars is considered to be the PWP. Of course some plants can pull harder than others, so -15 bars is the PWP by definition, not because any one specific plant will wilt at that point. Typical qv values for soils at the PWP are:

- Sandy ~ 0.07
- Loamy ~ 0.14
- Clay ~ 0.21

Water held more loosely than FC drains away by gravity. Water held more tightly than PWP is generally unavailable to plants. Water between PWP and FC is plant available water PAW. Not all PAW is used by plants; some evaporates.

Each soil has a unique available water capacity. Available water capacity (AWC) is the maximum PAW a soil could possibly provide. AWC tells the size of the container, PAW tells how full it is. AWC = qv at FC - qv at PWP. Typical AWC values for various soils are as follows:

- For sand ≈ 0.15 - 0.07 = 0.08
- For loam ≈ 0.30 - 0.14 = 0.16

  For clay ≈ 0.40 - 0.21 = 0.19

Technically, the term is unitless, but can be thought of 0.08 inches of water per inch of soil, or 0.08 meters of water per meter of soil, etc. AWC is reported in soil surveys as inches per inch.

Soils vary greatly in AWC. For example, within one Texas county:

- Clay AWC values range from about 0.06 to 0.20 in/in
- Loams range from about 0.03 to 0.20
- Sands range from about 0.02 to 0.10.

Infiltration, Run-off and Redistribution of Soil Water

Infilration

Infiltration refers to the movement of water into the soil layer. The rate of this movement is called the infiltration rate. If rainfall intensity is greater than the infiltration rate, water will accumulate on the surface and run-off will begin.

Movement of water into the soil is controlled by gravity, capillary action, and soil porosity. Of these factors soil porosity is most important. A soil's porosity is controlled by its texture, structure, and organic content. Coarse textured soils have larger pores and fissures than fine-grained soils and therefore allow for more water flow. Pores and fissures found in soils can be made larger through a number of factors that enhance internal soil structure.

For example, the burrowing of worms and other organisms and penetration of plant roots can increase the size and number of macro and micro-channels within the soil. The amount of decayed organic matter found at the soil surface can also enhance infiltration. Organic matter is generally more porous than mineral soil particles and can hold much greater quantities of water.

The rate of infiltration normally declines rapidly during the early part of a rainstorm event and reaches a constant value after several hours of rainfall. A number of factors are responsible for this phenomena, including:

- The filling of fine soil pores with water reduces capillary forces.
- As the soil moistens, clay particles swell and reduce the size of pores.
- Raindrop impact breaks up soil clumps, splashing fine particles into pores.

Soil Water Storage

Within the soil system, the storage of water is influenced by several different forces. The strongest force is the molecular force of elements and compounds found on the surface of soil minerals. The water retained by this force is called hygroscopic water and it consists of the water held within 0.0002 millimeters of the surface of soil particles. The maximum limit of this water around a soil particle is known as the hygroscopic coefficient. Hygroscopic water is essentially non-mobile and can only be removed from the soil through heating. Matric force holds soil water from 0.0002 to 0.06 millimeters from the surface of soil particles. This force is due to two processes: soil particle surface molecular attraction (adhesion and absorption) to water and the cohesion that water molecules have to each other. This force declines in strength with distance from the soil particle. The force becomes non-existent past 0.06 millimeters. Capillary action moves this water from areas where the matric force is low to areas where it is high. Because this water is primarily moved by capillary action, scientists commonly refer to it as capillary water. Plants can use most of this water by way of capillary action until the soil wilting point is reached. Water in excess of capillary and hygroscopic water is called gravitational water. Gravitational water is found beyond 0.06 millimeters from the surface of soil particles and it moves freely under the effect of gravity. When gravitational water has drained away the amount of water that remains is called the soil's field capacity.

Water Run-off from Soil

Aggregates of soil particles whose formation has not been influenced by human intervention are called peds. The peds in the surface horizons of soils develop into clods under the effects of cultivation and the traffic of urbanization. Soils whose A horizon is dense and unstructured increase the fraction of precipitation that will become surface run-off and have a high potential for erosion and flooding. These soils include not only those whose peds have been degraded but also coarse-textured soils with low porosity, particularly those of arid regions.

A well-developed clay horizon (Bt) presents a deep-lying obstacle to the downward percolation of water. Subsurface run-off cannot easily penetrate the clay layer and flows laterally along the horizon as it moves towards the stream system. This type of run-off is slower than its erosive counterpart over the land surface and leads to water saturation of the upper part of the soil profile and the possibility of gravity-induced mass movement on hillslopes (*e.g.*, landslides). It is also responsible for the translocation (migration) of dissolved products of chemical weathering down a hillslope sequence of related soil profiles (a toposequence). Subsurface water flow is also influenced by macropores, which, as noted above, are created through plant root growth and decay, animal burrowing activities, soil shrinkage while drying, or fracturing. In general, subsurface run-off processes are characteristic of soils in humid regions, whereas surface run-off is characteristic of arid regions and, of course, any landscape altered significantly by cultivation or urbanization.

Surface Run-off

Surface run-off is water, from rain, snowmelt, or other sources, that flows over the land surface, and is a major component of the water cycle.Run-off that occurs on surfaces before reaching a channel is also called overland flow. A land area which produces run-off draining to a common point is called a watershed. When run-off flows along the ground, it can pick up soil contaminants such as petroleum, pesticides, or fertilizers that become discharge or overland.

Surface run-off may be generated in three ways:

When the rate of rainfall on a surface exceeds the rate at which water can infiltrate the ground, and the depression storage has been filled. This is called infiltration excess overland flow, Hortonian overland flow, or unsaturated overland flow.

This more commonly occurs in arid and semi-arid regions, where rainfall intensities are high and the soil infiltration capacity is reduced because of surface sealing, or in paved areas.

When the soil is saturated and the depression storage filled, and rain continues to fall, the rainfall will immediately produce surface run-off.This is saturation excess overland flow or saturated overland flow.

After water infiltres the soil on an up-slope portion of a hill, the water may flow laterally through the soil, and exfiltrate (flow out of the soil) closer to a channel. This is called subsurface return flow or interflow.

As it flows, the amount of run-off may be reduced in a number of ways: a small portion of it may evaporate; water may become temporarily stored in microtopographic depressions; and a portion of it may become run-on, which is the infiltration of run-off as it flows overland. What surface run-off that remains eventually flows into a receiving water body such as a river, lake, estuary or ocean

Human Impact on Surface Run-off

Urbanization increases surface run-off, by creating more impervious surfaces such as pavement and buildings do not allow percolation of the water down through the soil to the aquifer. It is instead forced directly into streams, where erosion and siltation can be major problems, even when flooding is not. Increased run-off reduces groundwater recharge, thus lowering the water table and making droughts worse, especially for farmers and others who depend on water wells.

When anthropogenic contaminants are dissolved or suspended in run-off, the human impact is expanded to create water pollution. This pollutant load can reach various receiving waters such as streams, rivers, lakes, estuaries and oceans with resultant water chemistry changes to these water systems and their related ecosystems.

A contrarian could note there is considerable surface run-off in natural systems from animal wastes being entrained in run-off or from natural sediment loading in the absence of human alteration of the land. While these statements are true, they fail to convey that the most pernicious consequences to human health and ecosystems are from run-off issues related human intervention; however, in underdeveloped countries the proportion of run-off attributable to natural factors has greater dominance, principally due to the lack of isolation of water supplies from potential animal waste carrying run-off.

Effects of Surface Run-off

Erosion

Surface run-off is one of the causes of erosion of the earth's surface. Reduced crop productivity usually results from erosion, and these effects are studied in the field of soil conservation. There are four principal types of erosion: splash erosion, gully erosion, sheet erosion and streambed erosion. Splash erosion is the result of mechanical collision of a raindrop with the soil surface, resulting in some of the soil particles becoming suspended in surface run-off solution. Gully erosion occurs when the run-off stream is so strong that it cuts a recognizable swath in the soil and creates a small rivulet of well defined channel, which could be as small as one centimeter wide or as large as several meters. Sheet erosion is the overland transport of run-off which does not have any well defined channel. Both types of run-off can transport significant quantities of sediment or other water pollutants. In the case of gully erosion, massive amounts of material can be transported in a small time increment. Streambed erosion is not the subject of this article, but is the attrition of streambanks or bottoms by rapidly flowing rivers or creeks.

The soil particles carried in run-off have variable size typically from about.001 millimeter to 1.0 millimeter in diameter. Larger particles tend to precipitate out, or settle, with small transport distances, whereas minute particles can travel long distances suspended in the water column. For this reason it is the smaller particle silty soils that tend to generate turbidity and diminish light transmission, a condition that can disrupt aquatic ecosystems.

One of the main sources of erosive soil loss worldwide in the year 2006 stems from slash and burn treatment of tropical forests. When the total ground surface is stripped of vegetation and then seared of all living organisms, the upper soils are vulnerable to both wind and water erosion. In a number of regions of the earth, entire sectors of a country have been rendered unproductive; for example, on the Madagascar high central plateau, comprising approximately ten per cent of that country's land area, virtually the entire landscape is sterile of vegetation, with gully erosive furrows typically in excess of 50 meters deep and one kilometer wide. Shifting cultivation is a farming system which sometimes incorporates the slash and burn method in some regions of the world.

Environmental Impacts

The principal environmental issues associated with run-off are the impacts to surfacts water, groundwater and soil through transport of water pollutants to these systems.

Ultimately these consequences translate into human health risk, ecosystem distubance and aesthetic impact to water resources. Some of the contaminants that create the greatest impact to surface waters arising from run-off are petroleum substances, herbicides and fertilizers. Quantitative uptake by surface run-off of pesticides and other contaminants has been studied since the 1960s, and early on contact of pesticides with water was known to enhance phytotoxicity[5].

In the case of surface waters, the impacts translate to water pollution, since the streams and rivers have received run-off carrying various chemicals or sediments. When surface waters are used as potable water supplies, they can be compromised regarding health risks and drinking water aesthetics (that is, odour, colour and turbidity effects).

Contaminated surface waters risk altering the metabolic processes of the aquatic species that they host; these alterations can lead to death, such as fish kills, or alter the balance of populations present. Other specific impacts are on animal mating, spawning, egg and larvae viability, juvenile survival and plant productivity.

In the case of groundwater, the main issue is contamination of drinking water, if the aquifer is abstracted for human use. Regarding soil contamination, run-off waters can have two important pathways of concern. Firstly, run-off water can extract soil contaminants and carry them in the form of water pollution to even more sensitive aquatic habitats. Secondly, run-off can deposit contaminants on relatively pristine soils, creating health or ecological consequences.

Flooding

Flooding occurs when a watercourse is unable to convey the quantity of run-off flowing downstream. The frequency with which this occurs is described by a return period. Flooding is a natural process, which maintains ecosystem composition and processes, but it can also be altered by land use changes such as river engineering.

Floods can be both beneficial to societies or cause damage. Agriculture along the Nile floodplain took advantage of the seasonal flooding that deposited nutrients beneficial for crops.

However, as the number and susceptability of settlements increase, flooding increasingly becomes a natural hazard. Adverse impacts span loss of life, property damage, contamination of water supplies, loss of crops, and social dislocation and temporary homelessness. Floods are among the most devastating of natural disasters.

Agricultural Issues

A common context of run-off deals with agriculture. When farmland is tilled and bare soil is revealed, rainwater carries billions of tons of topsoil into waterways each year, causing loss of valuable topsoil and adding sediment to produce turbidity in surface waters.

The other context of agricultural issues involves the transport of agricultural chemicals (nitrates, phosphates, pesticides, herbiciedes etc) via surface run-off. This result occurs when chemical use is excessive or poorly timed with respect to high precipitation. The resulting contaminated run-off represents not only a waste of agricultural chemicals, but also an environmental threat to downstream ecosystems. The alternative to conventional farming is organic farming which eliminates or greatly reduces chemical usage.

Measurement and Mathematical Modeling

Run-off is analyzed by using mathematical models in combination with various water quality sampling methods. Measurements can be made using continuous automated water quality analysis instruments targeted on pollutants such as specific organic or inorganic chemicals, pH, turbidity etc. or targeted on secondary indicators such as dissolved oxygen. Measurements can also be made in batch form by extracting a single water sample and conducting any number of chemical or physical tests on that sample.

Computer model formed the basis of much of the mitigation study that led to strategies for land use and chemical handling controls.

Other computer models have been developed (such as the DSSAM Model) that allow surface run-off to be tracked through a river course as reactive water pollutants. In this case the surface run-off may be considered to be a line source of water pollution to the receiving waters.

Mitigation and Treatment

- Mitigation of adverse impacts of run-off can take several forms:
- Land use development controls aimed at minimizing impervious surfaces in urban areas
- Erosion contols for farms and construction sites
- Flood control programmes
- Chemical use and handling controls in agriculture, landscape maintenance, industrial use etc

Regarding Land use controls, and others have encouraged research on methods of mimimizing total surface run-off by avoiding unnecessary hardscape. Many municpalities have produced guidelines and codes for land developers that encourage minimum width sidewalks, use of pavers set in earth for driveways and walkways and other design techniques to allow maximum water infiltration in urban settings.

Erosion controls have appeared since medieval times when farmers realized the importance of contour farming to protect soil resources. Beginning in the 1950s these agricultural methods became increasingly more sophisticated.

Flood control progams as early as the first half of the twentieth century became quantitative in predicting peak flows of riverine systems. Progressively strategies have been developed to minimize peak flows and also to reduce channel velocities. Some of the techniques commonly applied are: provision of holding ponds to buffer riverine peak flows, use of energy dissipators in channels to reduce stream velocity and land use controls (above) to minimize run-off.

Methods commonly applied are: requirements for double containment of underground storage tanks, registration of hazardous materials usage, reduction in numbers of allowed pesticides and herbicides and more stringent regulation of fertilizers and herbicides in landscape maintenance. In many industrial cases, pretreatment of wastes is required, to minimize escape of pollutants into sanitary or stormwater sewers.

Redistribution of Soil Water

The magnitude of hydraulic redistribution of soil water by roots and its impact on soil water balance were estimated by monitoring time courses of soil water status and multiple depths and root sap flow under drought conditions in a dry ponderosa ecosystem and in a moist Douglas- ecosystem. The fate of deurerated water applied to small plots to create a strong horizontal soil water potential gradient was also monitored to assess the potential for horizontal redistribution of water and utilization of redistributed water by co-occuring shallow-rooted plants. In a 20-year-old Douglas-fir stand, approximately 28 per cent of the water removed daily from the upper 2 m of soil was replaced by nocturnal hydraulic redistribution during late August. In an old-growth ponderosa pine stand, approximately 35 per cent of the total daily water utilization from the upper 2 m of soil appeared to be replaced by hydraulic redistribution during July and August. By late September, hydraulic redistribution in the ponderosa pine stand was no longer apparent, even though total water use from the upper 2 m of soil was nearly identical to that observed earlier. Based on these results, hydraulic redistribution would allow 21 and 16 additional days of stored water to remain in the upper soil horizons in the ponderosa pine and Douglas-fir stands, respectively, after a 60-day drought. At both sites, localized applications of deuterated water induced strong reversal of root sap flow and caused soil water content to cease declining or even temporarily increase at locations too distant from the site of water application to have been influenced by movement of water through the soil without facilitation by roots. Xylem water deuterium values of ponderosa pine seedlings suggested utilization of redistributed water. Therefore, hydraulic redistribution may enhance seedling survival and maintain overstory transpiration during summer drought. These first approximations of the extent of hydraulic redistribution in these ecosystem suggests that it is likely to be an important process in both wet and dry forests of the Pacific Northwest.

The physical process of hydraulic redistribution (HR) is driven by competing soil, tree and atmospheric water potential gradients, and may delay severe water stress for roots and other biota associated with the upper soil profile. We monitored soil moisture characteristics across three forest ecosystems in the Pacific Northwest to investigate the spatial and temporal variability of HR of water by roots. Soil water potential and volumetric water content were measured down to depths of 1 or 2 m, respectively, in an old growth (300+ years) and young (20 years) ponderosa pine stand, and in old growth (500+ years) and young (25 years) Douglas-fir stands. Significant diel fluctuations in soil water content and water potential indicated that HR occurred in all four ecosystems. Spatial variability of HR to the upper soil layers (20-60 cm) was large within each site. Initiation and continuance of HR were dependent upon reaching a threshold soil water potential of about -0.2 MPa (-0.05 to -0.6 MPa). Hydraulic redistribution occurred earlier in the season at the drier pine sites, and was induced at higher water potential thresholds than in the Douglas-fir stands.

In addition, we manipulated the system by irrigation, soil trenching, and tree removal to alter competing water sources and sinks. In the irrigation experiment, we applied 2100 l of 9000 l deuterated water to a 1 m2 plot over three weeks, and sampled soil and the surrounding vegetation for 36 days. Deuterated water was immediately taken up by the surrounding dominant 25 yr-old Douglas-fir trees and within 7 days was detected in surface soils in front of those trees, but at least 1 m from the watering site. By the end of the sampling period (36 days) deuterated water was detected in soils behind the target trees, and in Oregon grape, blueberries, and small understory hemlocks as much as 4 m away from the watering source. The amount of HR in the upper soil layers at the watering site was twice that of the control, and the amount of water utilized from the upper soil was also increased. Trenching was hypothesized to decrease HR by separating a section of soil from the trees that had access to ground water; but because a large root in the plot was severed from a nearby dominant Douglas-fir tree, HR and soil moisture actually increased. We assumed that once the large sink of the tree was removed from the root, the soil sink received more water from the root. To confirm this, we measured sapflow on several roots on another tree, and then cut down the tree. In some roots we found a dramatic reversal of flow from up the tree to flow out to the surrounding soil. These manipulations highlighted that the magnitude of HR is governed by a competition between shallow roots in dry soil and the aboveground portion of the tree for water taken up by deep roots. Hydraulically redistributed water can move several meters from the original source where it can be utilized by other vegetation.

Soil Water Retantion Relationalship

Water Retention

The pore space in soil is usually at least partly filled with water. When all pores are water-filled, the soil is said to be water-saturated. Unsaturated conditions occur when water is present only in the finer pores while the larger pores are air-filled. This phenomenon can be explained by considering the capillary processes. When capillaries of different sizes are placed in water, the water will rise to the highest level in the smallest capillary. The smaller the capillary, the greater the suction by which water is held. Stated in another way, the pressure head (*h*) is more negative in the smaller capillary. It is more difficult (requires more energy) to remove water from finer pores than from larger pores. It is important to know how strongly water is held by the soil at any given time, because this governs not only the rate of water movement but also the availability of water to plants.

The pressure head (or suction) can be best measured with a tensiometer, which in its most simple form consists of a water-filled porous cup that is in contact with the soil in Figure below. At first, the water level in the open end of the tensiometer corresponds with the level of the cup. Gradually, however, water will be drawn by the unsaturated soil. This becomes ever more difficult as the level in the open end of the tensiometer decreases.

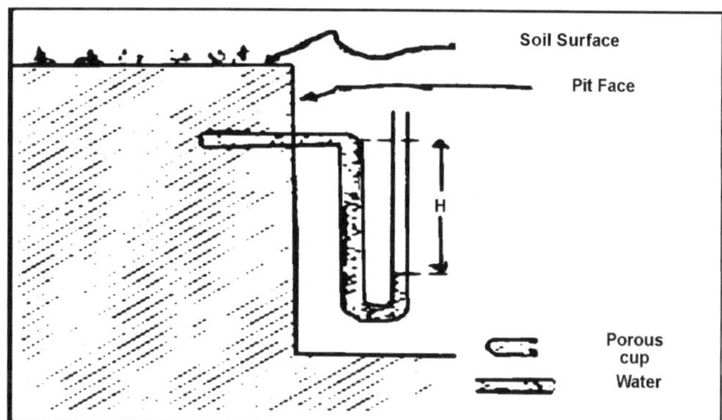

At Schematic diagram of a tensiometer in its simplest form. At equilibrium, the soil water around the porous cup has a pressure of -h centimeters (or inches). In practice, tensiometers used in the field are more elaborate than the one depicted here. Ordinarily, pressures can be read on the vertical manometer arm of a dial-type pressure guage above the ground surface.

equilibrium this level is a measure of the average

energy status of the water in the soil, which can be expressed as a negative pressure head of *h* centimeters (or inches). Why is the pressure head negative? If water with atmospheric pressure ($h=Ocm$) is brought into contact with unsaturated soil, it is drawn into the soil which acts like a sponge. Since water flow occurs from areas of high to low energy, we conclude that the pressure head in unsaturated soil is negative.

These principles are not only used to measure actual energy conditions, with tensiometers, but also to produce them artificially in soil samples to determine how much water can be held in different soils at specific energy levels. Differences among soils are likely because of different pore size distributions.

The apparatus used (in its simplest form) consists of a porous-plate which can be subjected to a series of suctions (negative pressure heads). Saturated soil samples can be placed on the plate and the soil-water content can be measured at several different suctions. Different points of a soil-water characteristic curve are thus obtained in Figure. Water content at each of the several suctions is obtained by removing a soil sample from the porous-plate apparatus and then weighing the sample before and after oven-drying to determine how much water has been retained by the soil at the suction in question.

Figure. Soil-water characteristic curves for surface soil and subsoil of Lakeland and Orangeburg soils. Notice that at any given pressure head (suction), the four different samples of soil retain markedly different amounts of water.

Soils with large pores can retain less water at a given suction than soils with smaller pores. This is because the smaller pores act as narrower capillaries, exerting greater suction on soil water than can be exerted by larger pores.

To illustrate the relationships among pore size, pressure head, and water retention, consider the soil-water characteristic curves in Figure above. At a pressure head of 0 cm, all or nearly all of the pores in each soil are filled with water, and the water contents of the four soils are about the same (35-45 per cent). As the pressure head is reduced (becomes more negative), however, the different soils exhibit different water retention behaviour. The Lakeland subsoil, for example, whose pores are predominantly large, has a water content of less than 10 per cent at a pressure head of -100 cm. The Organgeburg subsoil, by contrast, is much more clayey and has much finer pores than the Lakeland subsoil. At a pressure head of -100 cm, the Orangeburg subsoil has a water content greater than 30 per cent.

Energy Gradients

Water in soil moves from points where it has a relatively high energy status to points where its energy status is lower. Two factors determine the energy status of water at any given point in a non-saline soil. The first factor is the elevational position in the soil relative to a reference level. The higher an object is located above the reference level, the higher is its gravitational energy. This is true also of any given quantity of water in soil: the higher the water is located in the soil profile, the higher is its gravitational energy (g). Gravitational energy is expressed as the number of centimeters (or inches) above or below an arbitrarily chosen reference level.

The second factor determining the energy status of water is the water's pressure head (h), as discussed in Part II. Pressure heads are negative in unsaturated soil where not all the pores are water-filled. Those that are water-filled act in a manner similar to capillaries and exert suction on water. On the other hand, pressure head can be positive, as occurs in the saturated zone below the surface of the water in soil (the "water table"). There, all or nearly all (allowing for some trapped air) pores are filled with water. Pressure head is equal to zero ($h = 0$) at the water table. It becomes progressively higher (more positive) with increasing distance below the water table. Pressure head becomes progressively lower (more negative) with increasing distance above the water table. The hydraulic head (H) at a given point in the soil is equal to the sum of the gravitational and pressure heads, at that point, as illustrated schematically in Figure for selected conditions.

Figure. Schematic diagrams indicating the determination of pressure head (h), graviational head (g), and total head ($H = h|g$) at different points in the soil. The reference level for determination of g is taken to coincide with the elevation of the lower tensiometer cup. In part A, because H is higher at the higher cup, downward flow of water will occur provided that the soil is permiable. In part B, no flow is indicated because H is the same at both ends.

Water Movement

Direction and rate of water flow depend on differences in hydraulic head between different points in the soil. Not only is head difference important between two points, the distance between the points is also important. Expressed in terms of weather phenomena, we may observe a gentle breeze when high and low pressure areas are far removed from each other. When the same high and low pressure areas are very close to each other, however, much stronger winds prevail. So too with water flow in soils. The hydraulic head difference between two points in the soil must be divided by the distance between those points (L) to obtain the "gradient" of the hydraulic head between the two points. Direction of flow will be from a zone of higher H to a zone of lower H.

In order for water actually to move from one point to another, however, two conditions must be met. First, there must be a difference in hydraulic head between the two points (that is, *delta H* must be greater or less than zero). Second, the soil between these two points must be permeable enough to allow the movement of water. "Hydraulic conductivity" (K) is a measure of this ability of a soil to transmit water. The larger the K of a soil, the greater will be the movement of water through it for any given hydraulic gradient.

"Darcy's Law" for liquid movement in porous media states that the rate of water flow (q) through a given soil segment is equal to the hydraulic conductivity of that soil multiplied by the hydraulic gradient that exists in that soil. Darcy's Law is written mathematically as follows: q = -(*delta H/L*)K, where q = flux, or flow rate in centimeters per hour or day (or volume in cubic centimeters per square centimeter of soil area per hour or day); K = hydraulic conductivity in centimeters per hour or day; *delta H* = hydraulic head difference between two points in centimeters (which is composed of the gravitational and the pressure heads); and L = distance between the two points in centimeters.

A soil has a maximum K value when saturated (K_{sat}) and another, lower, K value for each lower water content and corresponding negative pressure head. Such values are characteristically different for different soils, depending upon soil structure and pore-size distribution. Even though the sand is more permeable than the clay at saturation,

In fact, water will only flow into the larger pores when the finer pores cannot conduct the applied water at the particular rate of application. Soil-water characteristic curves and hydraulic conductivity curves are needed to describe flow of water in soils. Flow rates of water in unsaturated soil cannot, unfortunately, be measured directly with a flow meter. The practical procedure to be followed involves the measurement of the pressure head at several points in the soil to determine the hydraulic gradient (*delta H/L*). A *K-h* curve is used to obtain the appropriate *K* value. The flow rate of the water (*q*) is found by multiplying *K* by (*delta H/L*).

The direction (upward, downward, or lateral) and magnitude of water flow in soils therefore depends upon the direction and magnitude of hydraulic head gradient and the degree of water saturation of the soil. Thus there can be no flow of water (*q*) in soil without both a hydraulic gradient (*delta H/L*) and hydraulic conductivity (*K*). A soil having a very high hydraulic conductivity will experience little water movement if there is a very low hydraulic gradient. On the other hand, a high hydraulic gradient between two points in the soil will not cause water flow if *K* is essentially zero due to the occurrence of impermeable soil between the two points.

Evaporation

Evaporation is one of the two forms of vaporization. It is the process whereby atoms or molecules in a liquid state (or solid state if the substance sublimes) gain sufficient energy to enter the gaseous state. It's the opposite process of condensation.

The thermal motion of a molecule of liquid must be sufficient to overcome the surface tension and evaporate, that is, its kinetic energy must exceed the work function of cohesion at the surface. Evaporation therefore, proceeds more quickly at higher temperature, at higher flow rates between the gaseous and liquid phase and in liquids with lower surface tension (*i.e.* higher vapor pressure). Since only a small proportion of the molecules are located near the surface and are moving in the proper direction to escape at any given instant, the rate of evaporation is limited. Also, as the faster-moving molecules escape, the remaining molecules have lower average kinetic energy, and the temperature of the liquid thus decreases. This phenomenon is also called evaporative cooling. One example for evaporative cooling is sweating.

Gas has less order than liquid or solid matter, and thus the entropy of the system is increased, which always requires energy input. This means that the enthalpy change for evaporation ($\ddot{A}H_{evaporation}$) is always positive. Evaporation is a cooling process.

Evaporation is a critical component of the water cycle, which is responsible for clouds and rain. Solar energy drives evaporation of water from oceans, lakes, moisture in the soil, and other sources of water. In hydrology, evaporation and transpiration (which involves evaporation within plant stomata) are collectively termed evapotranspiration.

Evaporative Equilibrium

If the evaporation takes place in a closed vessel, the escaping molecules accumulate as a vapour above the liquid. Many of the molecules return to the liquid, with returning molecules becoming more frequent as the density and pressure of the vapour increases. When the process of escape and return reaches an equilibrium, the vapour is said to be "saturated," and no further change in either vapor pressure and density or liquid temperature will occur.

Evaporation of Water

Evaporation is the process where a liquid, in this case water changes from a liquid to a gaseous state. It is a misconception that at 1 atm, water vapor *only* exists at 100 °C. Water molecules are in a constant state of evaporation and condensation flux near the surface of liquid water.

If a *surface* molecule receives enough energy, it will leave the liquid and turn into vapour pending an allowable vapor pressure. Under a pressure of 1 atm, water will boil at 100 °C.

The ratio of the heat loss from a pond by evaporation to the heat loss due to convection, independent of windspeed, is given by:

where $Q_{c,pa}$ is the heat loss from the pond by convection, in W/(m^2·K), Q_e is the heat loss from the pond by evaporation, in W/(m^2·K), T_p and T_a are the Kelvin (or Celsius) temperatures of the water and air, and P_{wp} and P_a are the vapor pressures of the pond surface and air, and p is the barometric pressure, with all pressures in mm Hg. (Bowen, 1926)

Factors Influencing Rate of Evaporation

Concentration of the substance evaporating in the air. If the air already has a high concentration of the substance evaporating, then the given substance will evaporate more slowly.

Concentration of other substances in the air. If the air is already saturated with other substances, it can have a lower capacity for the substance evaporating.

Flow rate of air.

This is in part related to the concentration points above. If fresh air is moving over the substance all the time, then the concentration of the substance in the air is less likely to go up with time, thus encouraging faster evaporation.

In addition, molecules in motion have more energy than those at rest, and so the stronger the flow of air, the greater the evaporating power of the air molecules.

Concentration of other substances in the liquid. If the liquid contains other substances (such as salts), it will have a lower capacity for evaporation. This is due to Raoult's law.

Temperature of the substance. If the substance is hotter, then evaporation will be faster.

Inter-molecular forces. The stronger the forces keeping the molecules together in the liquid or solid state the more energy that must be input in order to evaporate them.

Applied Evaporation

Forced Evaporation

Forced evaporation or distillation is a process used in the separation of mixtures, in which a mixture is heated to drive off the more volatile component with a higher vapor pressure.

Combustion Vaporisation

The fuel droplets vaporize as they receive heat by mixing with the hot gases in the combustion chamber. Hcat(energy) can also be received by radiation from any hot refractory wall of the combustion chamber.

Film Deposition

Evaporation is a common method of thin film deposition used in industry. Sputtering tends to be slower compared to evaporative deposition. However plastic substrates typically cannot tolerate the bombardment with energetic neutral atoms that unavoidably occurs in a sputter chamber.

An important example of an evaporative process is the production of aluminized PET film packaging film in a roll-to-roll web system. Often, the aluminum layer in this material is not thick enough to be entirely opaque since a thinner layer can be deposited more cheaply than a thick one.

The main purpose of the aluminum is to isolate the product from the external environment by creating a barrier to the passage of light, oxygen, or water vapor.

For some applications, the fact that evaporative deposition proceeds in a line-of-sight fashion is an important advantage. The energy distribution of evaporated material tends to be Maxwellian, with a temperature derived from the evaporation source.

In contrast, the plasmas encountered in sputtering systems are far from thermal equilibrium and may have high-energy tails that contain particles with large random velocities.

Evaporation is therefore a gentler process with a better defined beam of source material that can be used to coat just one side of a substrate or even the side of etched surface features, as in MEMS processing.

The three main kinds of evaporation are thermal, electron-beam and resistive. In the thermal method, the crucible that holds the source material is radiatively heated by a filament that winds around it. In the electron-beam method, the current that heats the crucible is boiled off a filament and is attracted to the crucible by a high voltage.

Electron-beam evaporation is used with the highest melting elements. Resistive evaporation is accomplished by passing a large current through a wire or foil of the material that is to be deposited. Molecular beam epitaxy is a particularly sophisticated kind of thermal evaporation.

The Process of Evaporation

The process of evaporation in a closed container will proceed until there are as many molecules returning to the liquid as there are escaping. At this point the vapor is said to be saturated, and the pressure of that vapor (usually expressed in mmHg) is called the saturated vapor pressure.

Since the molecular kinetic energy is greater at higher temperature, more molecules can escape the surface and the saturated vapor pressure is correspondingly higher. If the liquid is open to the air, then the vapor pressure is seen as a partial pressure along with the other constituents of the air. The temperature at which the vapor pressure is equal to the atmospheric pressure is called the boiling point.

Soil Temperature

Soil temperature plays an important role in many processes, which take place in the soil such as chemical reactions and biological interactions. Soil temperature varies in response to exchange processes that take place primarily through the soil surface. These effects are propagated into the soil profile by transport processes and are influenced by such things as the specific heat capacity, thermal conductivity and thermal diffusivity.

Thermocouple thermometers were placed at depths 1cm, 10cm and 25cm in one plot. In the other plot thermocouples were placed at 1cm and 10 cm. The thermocouple thermometers were recorded every 10 minutes throughout the experiment. Three mercury thermometers were also used to test the accuracy of the thermocouples.

The thermometers were placed at 1cm, 10 cm and in the air in the un-mulched plot. The experiment was run for 24 hrs recording temperature to check that every thing was working correctly, then mulch was added to the plot containing only two electrodes. And the experiment was run for another two days to test the effect of the mulch. The mercury thermometers were read every hour during the time that we worked in the field. A thermocouple is a combination of two metals, that produce a differing resistance that is temperature dependant. A common thermocouple used today is a combination of Copper and Canstantan. With proper calibration this electrode is accurate in the temperature range -262 C – 350 C. Copper of high electrical conductivity and oxygen content is very homogenous, and gives a highly reproducible thermoelectric power against platinum.

The constantan used with iron for thermocouples cannot be employed with copper, because of different voltage requirements. It is therefore necessary to use a constantan especially formulated for copper. This alloy is known as Adams Constantan and is not specified as a definite composition, but is any copper-nickel alloy that, combined with copper conforms to ASTM B3-45, matches the Adams copper-constantan table.

It is important to calibrate the electrode to temperature. The EMF produced from the electrode increases linearly with increases with temperature. In the field temperature range, the thermocouple would increase approximately 40 V per degree Celsius that the temperature raises.

'The simplest mathematical representation of the fluctuating thermal regime in a soil profile, is to assume that at all depths in the soil the temperature oscillates as a pure harmonic (sinusoidal) function of time around an average value' (Hillel, 1980). At each succeeding depth, the peak temperature is dampened and shifted progressively in time. The degree of damping increases with depth and is related to the thermal properties of the soil and the frequency of the temperature fluctuation.

Plots of temperature verses time were fitted with a sinusoidal function for depths of 1cm, 10cm and 25cm.

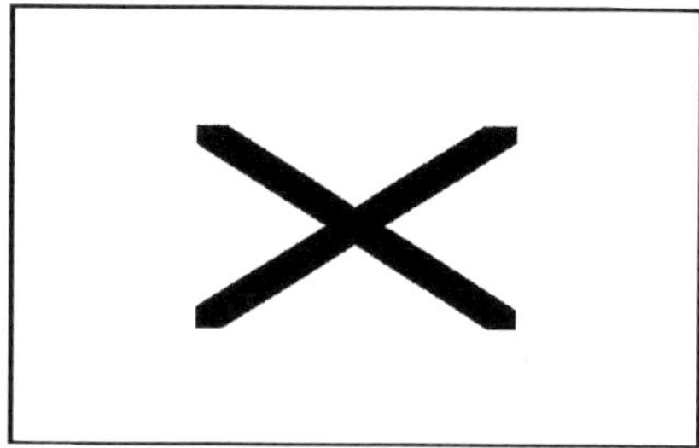

The highest peak at 42 degrees is the temperature at 1cm, the second highest peak is the temperature at 10cm and the lowest amplitude is the temperature at 25cm below the soil surface. This data clearly shows how damping increases with depth.

Modifying the soils thermal regime has great advantages for agricultural practices. These include controlling germination and plant growth through increasing or decreasing the soils temperature. This is most effectively done by manipulating the soil surface. To do this such things as mulches or other coverings as well as tillage practices may be used to warm or cool the soil. The overall effect is to modify the temperature amplitude at the soil surface and consequently throughout the profile. Mulches, depending on what they are and their albedo, will either warmer or cool the soil. A highly reflective material will lower the temperature by reducing the radiant flux reaching the soil. A more dense and less reflective material will increase the soil temperature by inhibiting evaporation.

At Narabri, a mulch consisting of straw was applied to the soil surface and the effects were monitored. The effects of the mulch include:

- Initially the mulch had an instaneous affect on the 1cm level of soil, the mulch had an instantaneous cooling effect.
- The mulch acts to remove reduce the amplitude (*i.e.* knocks out peaks and trouphs).
- The low thermal conductivity of the mulch reduces heat transferr over the surface of the soil, giving a damping effect.
- By the second day after the addittion of the mulch, the mulch reduced the amplitude of the 1cm by 17C and 10 cm by 7C.

Soil Temperature for Germination

Soil temperature can serve as a useful guide for timing spring seeding operations. Once the seed is planted, it is important that germination proceed uniformly and without delay. If the soil is too cool, germination is delayed, which results in seed damage and uneven or inadequate seeding emergence. It may even be necessary to reseed. On the other hand, if the soil is unusually warm in the spring, it may be better to seed earlier than normal. In either case, the soil temperature at seeding depth can be a practical guide.

Measuring Soil Temperature

Instruments

Any thermometer that will measure temperature at a specific depth can be used to measure soil temperature. This category includes ordinary glass bulb thermometers, provided they are long enough to permit reading the temperature when the bulb is inserted to the required depth.

Thermometers with a temperature scale etched onto the glass itself can be used more effectively than those with an attached scale. If the temperature scale is not etched onto the glass, a permanent marker can be used to mark in at least a few key values. The glass part of the thermometer should then be separated from the manufacturer's scale for taking readings.

Care must be taken to avoid breaking glass thermometers. Before inserting the thermometer into the soil, make a pilot hole in the soil using a rod, screwdriver, etc. of similar diameter as a precaution.

A Representative Site

Sites selected for temperature measurement should be representative of the land to be seeded in terms of the following factors:

Soil cover: whether stubble, summerfallow, strawcovered, bare, etc. Soil warms from the top down. Bare soil warms most quickly.

Aspect (amount and direction of slope): South-facing slopes warm more quickly than north-facing slopes or level land; snow also melts earlier on these southern slopes.

Soil moisture: Dry soils warm faster than wet ones; moisture content is strongly related to soil cover, aspect, texture and drainage.

Taking Readings

Place the bulb or sensing part of the thermometer at the intended seeding depth. In the case of glass thermometers, use a pilot hole to reduce the risk of breakage. The bulb should be in firm contact with the surrounding soil.

Allow a minute or two for the thermometer and the soil to reach temperature equilibrium before taking a reading. If there is bright sunshine, shade the thermometer during this time interval.

Record the Temperature.

Table Column1 and 2 give minimum and preferred temperatures established for a number of crops. These values should be regarded as approximate since germination depends on several factors.

For most crops, there is a temperature range just above the minimum germination temperature where the rate of germination is reduced. Spring seeding should not begin until soil temperature at seeding depth reaches or exceeds the required minimum. Shallow depths warm up first. Therefore shallow seeding gives the earliest germination, provided moisture requirements are met.

Table. Germination Temperatures for Field Crops.

Cereals and Oilseeds	Minimum (°C)	Preferred (°C)
wheat	4	20
barley	3-5	20
oats	5	20-24
canola/rapeseed	5 (Argentine) 7-10 (Polish)	15-20 15-20
Forage Crops		
alfalfa	1	25
birdsfoot trefoil	1	26
red clover	3	25
sweet clover	1	18-25
white clovers	5	18-20
fescues	3	13-18
orchardgrass	4	18-20
timothy	4	18-22

Table. Germination Temperatures for Vegetables and Special.

Crops	Minimum (°C)	Preferred (°C)
bean	8-10	16-30
beet	4	10-30
cabbage	4	7-35
carrot	4	7-30
cauliflower	4	7-30
celery	4	15-21
corn	10	16-32
cucumber	16	16-35
eggplant	16	24-32
lettuce	2	4-27
onion	2	10-35
parsley	4	10-30
parsnip	2	10-21
pea	4	4-24
pepper	16	18-35
pumpkin	16	21-32
radish	4	7-32
rutabaga	4	16-30
spinach	2	7-24
squash	16	21-35
tomato	10	16-30

Table. Average Monthly Soil Temperature at 5 cm (°C).

	April		May	
	AM	PM	AM	PM
Fort Vermilion	0.2	2.5	8.6	15.8
Peace River	1.3	4.6	8.2	14.2
Beaverlodge	0.3	2.4	6.6	11.9
Vegreville	0.8	3.6	7.2	12.1
Calgary	1.2	6.1	7.1	14.6
Vauxhall	3.2	8.9	10.4	17.5
Lethbridge	4.0	7.9	10.5	16.0

8

Soil and Water Conservation on Agricultural Land

Broadly speaking, the practical methods of soil and water conservation fall into two important classes, *viz.* Agronomic measures and mechanicla measures.

SOIL AND WATER-CONSERVATION MEASURES

The key to soil and water conservation is the utilization and treatment of land according to its capability.

LAND-CAPABILITY CLASSIFICATION

Any soil and water-conservation project includes two distinct sets of operations, viz.

- The mapping of land for classification according to its capability,
- Planning and executing measures to check erosion, improve land productivity and reclaim wasteland.

The farm plans for effective soil and water conservation are based largely on the capability of the land. The land-capability classification map is normally prepared by interpreting a standard soil-survey map.

Land-capability classification is a systematic arrangement of different kinds of lands according to those properties that determine the ability of the land to produce crops on a virtually permanent basis.

The Factors Determining Land-capability

These are the major soil characteristics of the land, *e.g.,,* the texture of the top soil, its effective depth, permeability of the top soil and subsoil, and associated land features, *e.g.,,* the slope of the the land, the extent of erosion, the degree of wetness and susceptibility to overflowing and flooding.

The grouping of soils into capability classes is done primarily on the basis of their capability to produce common cultivated crops and pasture plants without deterioration over a long period.

Land-capability Classes

The land-capability classes are based on the intensity of hazards and the limitations of use. The land-capability classes range from the best and most easily farmed land to that which has no value for cultivation, grazing or forestry, but which may be suited to wild-life, recreation or for watershed protection. They all fall into 2 broad groups: one suitable for cultivation and other land uses, and the other not suitable for cultivation, but suitable for other land uses.

Land Suitable for Cultivation and Other Uses

Class I (Green Colour)

Soils in class I have very few or no limitations that restrict their use.

This type of land is nearly level and the erosion hazard is low. The soils are deep, well-drained, easily worked, hold water well and are either fairly well supplied with plant nutrients or highly responsive to the application of fertilizers. The soils are not subject to damage because of overflow. The local climate must be favourable for growing many of the common field crops. In irrigated areas, the soils may be in class I, if the limitation of the arid climate has been removed by relatively permanent irrigation works.

These soils need ordinary management practices to maintain productivity. Such practices may include the use of one or more of the following: fertilizers, lime, cover and green-manure crops, conservation of crop residues and crop rotations.

Soils in this class are suited to a wide range of plants, may be used for cultivated crops, pastures, forests, and wildlife, food and cover.

Class II (Yellow Colour)

Soils in class II have some limitations which reduce the choice of plants or require simple conservation practices.

The limitations of soils in class II may result from the effects of one or more of the following factors:

- A gentle slope,
- A slight susceptibility to erosion,
- Less than ideal soil depth,
- Occasional damaging overflow,
- Wetness which can be corrected by drainage, but existing permanently as a moderate limitation,
- Slight to moderate salinity or sodium, easily corrected but likely to recur,
- A slight climatic limitation on soil use and management.

These soils require careful management. The limitations are only a few and the practices are easy to apply. They may need one or more of the following practices: terracing, strip cropping, contour cultivation, water disposal area, covered with vegetation crop rotation, cover and green-manure crops, stubble mulching, the use of fertilizers, manure and lime. These soils may be used for growing cultivated crops, raising pastures, forests, and for wild-life, food and cover.

Class III (Red Colour)

Soils in class III have moderate limitations which reduce the choice of plants or require special conservation practices. Soils in class III have more restrictions than those in class II and, when used for cultivated crops, the conservation practices are usually more difficult to apply and to maintain.

Limitations of soils in class III may result from the effects of one or more of the following factors:

- A moderately sloping land,
- Moderately susceptibility to water or wind erosion,
- Frequent overflow accompanied with some crop damage,
- Very slow permeability of the sub-soil,
- Wetness or continuing water-logging after drainage,
- Shallow soil depth up to the bed-rock, hard-pan or clay-pan which limits the rooting-zone and water storage,
- Low moisture-holding capacity,
- Moderate salinity or sodium,
- Moderate climatic limitations.

Class IV (Blue Colour)

Soils in class IV have severe limitations that restrict the choice of plants and require careful management. The restrictions in the use of these soils are greater than those in class III and the choice of plants is more limited. When these soils are cultivated, very careful management is required and the conservation practices are more difficult to apply and to maintain.

The use of these soils for cultivated crops is limited as a result of the effect of one or more of the permanent features, such as:

- Steep slopes,
- Severe susceptibility to water or wind erosion,
- Severe effect of past erosion,
- Shallow soil,
- Low moisture-holding capacity,
- Frequent overflow accompanied with severe crop damage,
- Excessive wetness or continuing hazard of water-logging after drainage,
- Severe salinity or sodium,
- Moderately adverse climate.

These soils can be used for crops, pastures, forests, and wild-life food and cover.

Land Not Suitable for Cultivatio but Suitable for Other Land Uses

Class V (Dark Green or Uncoloured)

Soils in class V have little or no erosion hazard, but have other limitations, the removal of which is not practicable. They are used largely for pastures, forests, and wild-life food and cover.

Such land is nearly level and is not subject to more than slight wind or water erosion. Cultivation is not feasible because of one or more limitations, such as overflow, stoniness, wetness or severe climate.

Examples of class V land are:

- Soils of lowlands subject to frequent overflows which prevent the normal production of cultivated crops,
- Nearly level soils with growing season that prevents the normal production of cultivated crops,
- The level or nearly level stony or rocky soils,
- Ponded areas where drainage for cultivated crops is not feasible but where soils are suitable for grasses or trees.

Soils in class V are not suitable for raising cultivated crops, but are suitable for perennial vegetation (grazing and forestry, with few or no limitation). Pastures can be improved, and benefits from proper management can be expected. Physical conditions of soils are such that it is practicable to apply pasture improvements, if needed, such as seeding, liming, fertilizing, and water control with contour furrows, drainage ditches, diversions of water spreaders.

Class VI (Orange Colour)

Soils in class VI have severe limitations that make them unsuitable for cultivation and limit their use largely for pastures, or forests, or wild-life food and cover. Soils in class VI have continuing limitations which cannot be corrected, such as:

- Steep slope,
- Very severe erosion hazard,
- Very severe effect of past erosion
- Stoniness,
- Shallow rooting-zone,
- Excessive wetness or overflow,
- Low moisture capacity,
- Salinity or sodium,
- Severe climate.

Soils in this class are subject to moderate limitations under grazing or forestry use.

Class VII (Brown Colour)

Soils in class VII have very severe limitations that make them unsuitable for cultivation and restrict their use largely to grazing, or forestation, or wild-life food and cover. The soils in this class are subject to severe limitations or hazards under either grazing or forestry use.

The physical condition of soils is such that it is not practicable to adopt pasture improvements and water-control practices. Soil restrictions are sever e than those in the case of class VI soils.

Class VIII (Purple Colour)

Soils and land forms in class VIII have limitations that preclude their use for commercial plant production and restrict their use to recreation, wild-life food and cover or to water-supply, water shed protection or for aesthetic purposes.

Significant return on site benefits from soils and land forms in class VIII cannot be expected from management of crops, grasses or trees, although indirect benefits from wild-life, watershed protection or recreation may be possible.

Limitations which cannot be corrected may result from the effects of one or more of the following factors:

- Erosion or erosion hazard,
- Severe climate,
- Wet soil,
- Stones,
- Low moisture capacity,
- Salinity or sodium.

Bad lands, rock outcrops, sandy beaches, marshes, deserts, river wash, mine tailings and other nearly barren lands are included in class VIII in order to protect other more valuable soils to control water or for wild-life or for aesthetic reasons.

The land-capability class is indicated on the maps by roman numerals I to VIII or by standard colours or by both.

Land-capability Subclass

A land-capability class is determined by the degree of limitations in land use together with the hazards involved. For example, in class III land, we have land suitable for cultivation but subject to moderate hazard of water erosion because of a steep slope or a moderate hazard of wind erosion on smooth land or moderate hazard of water-logging or overflow and a shallow depth to the bed-rock. Each of these kinds of limitations are recognized at the subclass. Four kinds of limitations are recognized at the subclass level.

Subclass (e) Erosion

It is made up of soils where the susceptibility to erosion is the dominant hazard in their use. Susceptibility to erosion and damage due to past erosion are major factors that govern the placing of soils in this class.

Subclass (w) Excess Water

It is made up of soils where excess water is the dominant hazard in their use. Poor soil drainage, wetness, high water-table and overflow are the criteria for determining which soils belong in this subclass.

Subclass (s) Soil Limitations within the Rooting Zone

It includes soils which have such limitations as the shallowness of the rooting-zone, stones, low moisture-holding capacity, low fertility difficult to correct and salinity or alkalinity.

Subclass (c) Climatic Limitations

It includes soils where the climate (temperature or lack of moisture) is the only major hazard of limitations in their use. The land-capability subclass is designated by the small letters which follow the land-capability classes. Roman numerals, *e.g.,*, IIe, IIIs, IVw capability class I has no subclass. Where two kinds of limitation can be modified or corrected and are essentially equal, the subclasses have the following priority e, w, s, c. where soils have two kinds of limitations, the dominant one is shown first.

Table. Effect of Climate

Characteristics of Climate	**Influence on the Capability Group**
Humid climate with well-distributed rainfall	Can start with class I
Humid climate with occasional dry-spell	Can start with class II
Sub-humid, crop yield frequently reduced by droughts	Can start with class II or III
Semi-arid	Can start with class III or IV
Arid	Can start with class IV

AGRONOMIC MEASURES FOR SOIL AND WATER CONSERVATION

Agronomic practices for soil and water conservation help to intercept rain drops and reduce the splash effect, help to obtain a better intake of water rate by the soil by improving the content of organic matter and soil structure, help to retard and reduce the overland run-off through the use of contour cultivation, mulches, dense-growing crops, strip-cropping and mixed cropping.

CONTOUR-FARMING

During intense rain storms, the soil cannot absorb all the rain as it falls. The excess water flows down the slope under the influence of gravity. If farming is done up and down the slope, the flow of water is accelerated, because each furrow serves as a rill. The major part of the rain is drained away without infiltrating into the soil. The top fertile soil, along with plant nutrients and seeds, is washed off. All this results in a scanty and uneven growth of a crop. A simple practice of farming across the slope, keeping the same level, as far as possible (which is technically called contour-farming) has many beneficial effects. The ridges and the rows of the plants placed across the slope form a continual series of miniature barriers to the water moving over the soil surface. The barriers are small individually, but as they are large in number, their total effect is great in reducing run-off, soil erosion and loss of plant nutrients. It has been experimentally proved that contour-farming reduces run-off and prevents soil erosion as compared with the up-and-down cultivation in the major groups of soils in India, *viz.* Alluvial soils, black soils and deep lateritic soils.

Table. Effect of Contour Cultivation on Run-off and Soil Erosion

Treatment	Rainfall causing run-off (mm)	Run-off (mm)	Soil loss (tonnes/ha)
Alluvial soils - 8 per cent slope - Dehra Dun			
Maize (up-and-down cultivation)	1,223	670	28.5
Maize + cowpeas (contour cultivation)	1,223	511	19.3
Deep lateritic soil - 25 per cent slope - Ootacamund			
Potato (up-and-down cultivation)	1,295	52	39.3
Potato (contour cultivation)	1,295	29	14.9

Apart from conserving the water and soil, contour-farming conserves soil fertility and increases crop yields. Contour-farming on alluvial soil, 2.2 per cent slope at Kanpur has conserved 11.3 kg of N, 11.7 kg of P_2O_5, 44.4 kg of K_2O, 398.1 kg of CaO, 118.1.kg of MgO in one season alone. These nutrients converted into fertilizers amount to 56.5 kg of sulphate of ammonia, 70 kg of single super-phosphate and 74 kg of muriate of potash per hectare in one season.

Contour-farming has also given 490 kg of johar grain and 273 kg of johar stalks more than the up-and-down cultivation at Kanpur. Thus every mm of rain-water conserved by contour-farming gave 22.5 kg of johar grain and 12.5 kg of johar straw more than the up-and-down cultivation. It has been proved that much less power is required to be exerted by man, animals and machines, if the cultivation is done in the contour instead of up-and-down the slope. There is less wear and tear of the implements and the same job is done in less time when contour-farming is practiced.

How to Practice Contour-farming

On long slopes, bunding is usually done to reduce the length of the slope. These bunds will serve as a good guide for contour-farming. All the cultural operations have to be done parallel to these bunds.

On gentle slopes (between 0.5 to 2.0 per cent) bunding may not be essential. Contour (the line passing through the points having the same level) guidelines can be marked with the help of a hand level. On uniform slopes, these lines are to be marked about 50 m apart. Farming is done parallel to these lines.

The establishing of contour-farming on undulating land (having many depressions and ridges) is somewhat tedious. The water from each furrow collects in the depressions and results in breaches. The depressions are required to be filled up by levelling or may be left under grass.

Mulching

Surface mulches are used to prevent soil from blowing and being washed away, to reduce evaporation, to increase infiltration, to keep down weeds, to improve soil structure and eventually to increase crop yields. Inter-culture kills weeds and produces a five or seven cm thick soil mulch which helps to reduce evaporation from the top soil. It also breaks the surface crust which forms after each downpour. Studies on mulching carried out in India under rain-fed agriculture have concentrated on the measurement of crop responses rather than on the manner in which crop responses are influenced.

At Dehra Dun, mulching with maize residues did not significantly influence the yield of the succeeding yield crop.

At Kota, mulching, in general, gave a higher yield of wheat grain and stalk than no mulching. At Bellary, mulching with different materials (*e.g.,,* paddy husk, Encap Esso mulch, grass mulch, johar stubble) did not give encouraging results in increasing the yields of cotton and johar crops. Mulching with straw og Encap Esso mulch increased the infiltration rate, improved soil moisture and increased the yields of wheat, barley, gram and linseed, succeeding the maize crop on eroded soils at Ranchi.

Growing of Crops which Provide the Maximum Cover, Reduce Run-off and Soil Loss

Table. Soil and Water Loss Under Various Crops and Cropping Systems

Treatment	Run-off (mm)	Soil loss (tonnes/ha)
Alluvial soil - 2 per cent slope - rainfall (June to March) 791.1 mm rainfall causing run-off 385.9 mm - Vasad (average of 8 years		
Bajra-fallow	167.7	2.3
Bajra-mung strips (3:1 ratio)-fallow	121.0	2.2
Bidi tobacco-fallow	205.3	4.8
Sunhemp (cover-cum-green-manure crop) *bidi* tobacco-fallow	142.6	4.4
Black soil - 1 per cent slope - annual rainfall 656.8 mm - Kota (average of 4 years)		
Groundnut	73.9	1.9
Black-gram (*mash* or *urad*	102.6	2.1
Johar	103.3	2.4
Red soil - 2 per cent slope - monsoon rainfall 1,002 mm - Deochanda (DVC) (average of 3 years)		
Maize	64	3.33
Maize and *urad* (for seed) intercropped	64	3.0
Maize and *urad* (for green-manuring) intercropped	66	2.9
Maize and *arhar* intercropped	66	2.5
Maize and *arhar* (maize stalks or stubble mulched) - intercropped	86	3.3
Red soil - 5 per cent slope - monsoon rainfall 1,129 mm - Deochanda (DVC) (average of 3 years)		
Maize	326	23.6
Urad	418	47.0
Groundnut	325	13.9
Gora paddy	385	21.1

Cultivated legumes, in general, furnish a better cover and hence, better protection to cultivated land against erosion than ordinary cultivated crops. The crops and the cropping systems will naturally vary from region to region, depending on the soil and climatic conditions.

At Dehra Dun, cowpeas provided the maximum canopy with or without the application of phosphate fertilizer. This was followed by mung, urad and dhaincha.

At Vasad, cowpeas provided the best vegetative cover for the soils. Sunnhemp was the next best; mung and groundnut were effective to some extent.

At Kota, velvet bean formed the canopy effective to the maximum extent. Cowpea was the next best.

At Kanpur, mung served as the most effective canopy, followed by urad and guar. The application of phosphorus increased the canopy.

At Rehmankhera, cowpea formed the best canopy, followed by Styzolobium mung, groundnut, urad, moth, and soybean. Sunnhemp, mung and Styzolobium helped to conserve soil moisture well, leading to a higher yield in the case of the succeeding barley crop. All the legumes were equally effective in reducing the loss of soil and nutrients.

Thus among the legumes, cowpea and mung proved to be important crops for providing a good cover for the land during the rainy season.

Bidi tobacco, being a clean-cultivated crop, resulted in the maximum soil loss and water loss. The soil and water losses were very much reduced, especially the water loss, when a cover-cum-manure crop was grown during the early monsoon before tobacco was transplanted.

The inclusion of a cover-cum-green-manure crop increased the yield of tobacco from 1,585 to 1,850 kg/ha (*i.e.,*, an increase of 16.7 per cent) and reduced the need for nitrogenous fertilizer dose by 45 per cent. Bajra+mung strip-cropping also reduced the water loss considerably.

Groundnut grown under two soil-climate conditions has been shown to reduce both the soil and water loss.

Urad grown under two soil-climate conditions has been shown to be a run-off and soil-loss-permitting crop like maize. Intercropping maize with urad or arhar did not reduce the loss of soil and water.

STRIP-CROPPING

Strip-cropping is essentially another form of rotation. Its importance in controlling the run-off erosion and thereby maintaining the fertility of the soil is now universally recognized. Strip-cropping, in effect, employs several good farming practices, including crop rotation, contour cultivation, proper tillage stubble-mulching, cover-cropping, etc. Strip-cropping is of the following different forms:

- Contour strip-cropping,
- Field strip-cropping,
- Wind strip-cropping,
- Permanent or temporary buffer strip-cropping.

Contour Strip-cropping

Contour strip-cropping is the growing of soil-exposing and erosion-permitting crop in strips of suitable widths across the slopes on contour, alternating with strip of soil-protecting and erosion-resisting crops. Contour strip-cropping shortens the length of the slope, checks the movement of run-off water, helps to desilt it and increases the absorption of rain-water by the soil.

Further, the dense foliage of the erosion-resitant crops prevents the rain from beating the soil surface directly. It is advisable to rotate the strip-planting by sowing a non-resistant crop, following an erosion-resistant crop and vice versa.

Very little experimental work has been done to determine the best combination of crops and other factors in a strip-cropping programme.

The data collected at the Dry Farming Research Station, Sholapur, Maharashtra, have given the following indications:

- Groundnut, Moth bean (Phaseolus acontifolius) and Horse gram (Dolichos biflorus) are the most efficient and suitable crops for checking erosion.
- The normal seed-rates of leguminous crops, other than groundnut, do not give sufficiently dense canopies to prevent rain drops from beating the soil surface in much cases. The seed-rate should be trebled.
- The most effective width of the contour strips for cereals, such as jowar and bajra, is 21.6 m and for the intervening legume 7.2 m.

The results have also shown the following strip widths as suitable for erosion-permitting and erosion-resistant crops on different slopes:

Slope	Width of erosion-permitting crops	Width of erosion-resistant crops
1/2 per cent and below	45 m	9 m
Between 1 and 2 per cent	24 m	6 m
Between 2 and 3 per cent	13.5 m	4.5 m

Field Strip-Cropping

It is the planting of farm crops in more or less parallel strips across fairly uniform slopes, but not on exact contours.

Wind Strip-Cropping

It consists of planting tall-growing crops such as jowar, bajra or, maize, and low-growing crops in alternately arranged straight and long, but relatively narrow, parallel strips laid out right across the direction of the prevailing wind, regardless of the contour.

Permanent or Temporary Buffer Strip-cropping

In the case of permanent or temporary buffer strip-cropping, the strips are established to take care of critical, *i.e.,* steep or highly eroded, slopes in fields under contour strip-cropping. These strips do not form part of the rotation practiced in normal strip-cropping, and they are generally planted with perennial legumes, grasses or shrubs on a permanent or temporary basis.

MIXED CROPPING

The main reason for the lack of interest in strip-cropping in India is the small size of thc holdings of the farmers. Instead, the Indian farmers have been interested in mixed farming which is very extensively adopted by him. Some of the important objectives of mixed cropping are a better and continuous cover of the land, good protection against the beating action of the rain, almost a complete protection against soil erosion and the assurance of one or more crops to the farmer. The roots of various species in a mixed crop feed at different depths in the soil.

Improvements in the practice of mixed cropping tend towards the rationalization of the mixtures with respect to composition and content and towards better arrangements than from the usual medley of crops. The line-sowing of mixed crops gives rise to the practice of intercropping.

Table. Effects of Intercropping on the Yields of Crops - Black soil at Kota.

	Yield kg/ha (average of 5 years)
Wheat (pure crop)	314.9
Gram (pure crop)	314.6
Wheat + gram (in alternate rows)	439.7

Out of five years of trial, the wheat crop failed in one year owing to the extreme drought, and gram failed in one year because of the incidence of gram wilt. In both these years, the other mixed crops was a success, thus ensuring at least one crop to the farmer when mixed cropping was practiced. The wheat+gram mixture gave a consistently higher yield than the pure crops individually. The increase in net profit as a result of mixed cropping was 89 per cent and 183 per cent over the pure crops of wheat and gram, respectively.

Mechanical Measures of Erosion Control

Mechanical measures (also called engineering measures) play a very vital role in controlling erosion on agricultural land. They are adopted to supplement the agronomical practices when the latter alone are not adequately effective. These measures include basin-listing, subsoiling, contour-bunding (also called level terraces, ridge terraces, absorptive-type terraces), graded bunding and bench-terracing on steep slopes.

The main objectives of the mechanical measures for controlling erosion are:

- To increase the time of concentration by intercepting the run-off and thereby providing an oppurtunity for the infiltration of water, and
- To divide a long slope into several short ones so as to reduce the velocity of the run-off and thus prevent erosion.

Basin-Listing

Basin-listing consists in making of small interrupted basins along the counter with a special implement, called a basin-lister. Basin-listing helps to retain rain-water as it falls and is specially effective on retentive soils having mild slopes.

Subsoiling

This method consists in breaking with a subsoiler the hard and impermeable subsoil to conserve more rain-water by improving the physical conditions of a soil. This operation, which does not involve soil inversion and promotes greater moisture penetration into th soil, reduces both run-off and soil erosion. The subsoiler is worked through the soil at a depth of 30-60 cm at a spacing of 90-180 cm.

Contour-Bunding

This practice consists in making a comparatively narrow-based embankment at intervals across the slope of the land on a level that is along the contour. It is an important measure that conserves soil and water in arid and semi-arid areas with high infiltration and permeability, and is commonly adopted on agricultural land up to a slope of about 6 per cent. The cross-sections of bunds adopted in the dry areas of Maharashtra are given below.

These cross-sections of the bunds can be used as a guide in similar situations. It may be noted here that, as no specifications for bunding in deep black soil are available so far, no large-scale bunding on them can be recommended, pending further research.

As regards spacing between bunds it should be not exceed 150 cm vertical drop or 67.5 cm horizontal spacing, whichever is less.

Type of soil	Top width (cm)	Bottom width (cm)	Height (cm)	Side slope	Area of section (sq m)
Very shallow soils	45	195	75	1:1	10.00
Shallow soils	45	255	82.5	1.25:1	13.75
Medium soils	52.5	300	82.5	1.50:1	16.00
Medium heavy soils	60	420	90	2:1	24.00

The following schedule, which is used in Maharashtra, can be used as a fair guide:

Percentage slope	Vertical drop
0 to 1	1.05 m
1 to 1-1/2	1.20 m
1-1/2 to 2	1.35 m
2 to 3	1.50 m

Where the rainfall is more than 62.5 mm and moisture conservation is important, bunds may be constructed at shorter intervals. In the alluvial soils of Gujarat, a vertical interval of 1.83 m and a cross-section of 1.3 sq m were found to be suitable for land with slopes ranging from 6-12 per cent and for slopes less than 6 per cent contour bunds with cross-sections of 0.9 to 1.3 sq m spaced at 0.9 to 1.2 m vertical interval were found to be effective. Observations and experiments at Sholapur, Bellary and Kota have shown that even in semi-arid climate, contour-bunding in deep black soils, with montmorillonitic type of clay is not suitable. The low rates of permeability and infiltration in these soils cause a prolonged impounding of water on the upstream side, and the crops are consequently damaged.

Graded Bunding or Channel Terraces

Graded bunding has been used in areas receiving rainfall of more than 80 cm per year, irrespective of the soil texture. In clay soils, graded bunding has to be used even for areas having less than 80 cm of annual rainfall. Graded bunds may be narrow-based or broad-based. A broad-based graded terrace consists of a wide-low embankment constructed on the lower edge of the channel from which the soil is excavated. The channel is excavated at suitable intervals on a falling contour with a suitable longitudinal grade.

At Dehra Dun, studies on the channel terraces in the alluvial soils of the valley with 3-4 per cent slope have shown that the two vertical intervals (V.I.)* tried, *viz.* 0.3 (S + 2) and 0.3 (S + 3) did not show much difference in run-off, whereas the soil loss per mm of run-off was more from the V.I. = 0.3 (S + 3). However, the annual soil loss was within the permissable limits and, therefore, V.I. of 0.3 (S + 3) can be adopted safely. *V.I. = Vertical interval in metres. S = Slope per cent Frequency of run-off (cm) and soil loss (tonnes/ha) for two vertical intervals (Dehra Dun)

Vertical interval (m)	**Run-off (cm) Frequency (years)**				**Soil loss (tonnes/ha) Frequency (years)**			
	10	5	2	1.25	10	5	2	1.25
0.3 (S+2)	35.0	30.0	17.5	11.0	5.6	4.2	2.5	1.5
0.3 (S+3)	40.0	32.0	19.0	12.5	7.0	5.4	3.2	1.8

The channel grades of 0.4 per cent and 0.6 per cent did not show any signs of scouring.

Observations on cross-section revealed that the broad-based and narrow-based cross-sections functioned well without any special maintenance. Broad-based terracing is recommended where farming is practiced with tractors. Studies on broad-based channel terraces constructed on black soils (land slope, 1 per cent) in the semi-arid climate at Bellary showed that a V.I. of 0.6 m was better than that of 0.75 m and that the channel terrace with a variable grade showed less of run-off and soil loss than the channel terrace with a uniform grade. At Kota, graded bunds, 0.7 to 1.0 sq m cross-section and with 0.1 to 0.5 per cent channel grade were found suitable for clay loam to clay black soils.

Bench- Terracing

This consists of a series of platforms having suitable vertical drops along contours or on suitably graded lines across the general slope of the land. The vertical drop may vary from 60 to 180 cm, depending upon the slope andsoil conditions, as also on the economic width required for easy cultural operations. The material excavated from the upper part of the terrace is used in filling the lower part. A small 'shoulder' bund of about 30 cm in height is also constructed along the outer edge of the terrace. Bench terraces may be 'table top' or sloping outward or inward with or without a slight longitudinal grade, according to the rainfall of the tract - medium, poor or heavy, and the soil and the subsoil are fairly absorptive or poorly permeable. On steeply sloping and undulating land, intensive farming can be practiced only with bench-terracing. The initial cost of bench -terracing is more than that of bunding. However. Bench-terracing helps to retain the soil, moisture, manure and fertilizer better and facilitates the application of irrigation, if available. In rainfed areas, terracing is usually practiced on slopes ranging from 6 to 33 per cent. It may have to be used on gentle slopes, if irrigation is to be applied to the crop.

At Ootacamund, studies have been conducted on the optimum length of bench terraces, longitudinal and inward grades of bench terraces.

Effect of the length of Bench Terraces (on 13-16 per cent slope) on run-off, soil loss and the yields of crops - Ootacamund

Length of bench terrace (m)	91.5	122.0	152.0
Run-off (mm)	32.1	26.2	24.2
Run-off as percentage of the total rainfall	2.4	2.0	1.8
Soil loss (tonnes/ha)	0.59	0.55	0.77
Yield of potato (kg/ha)	10,047	9,418	9,692

The difference in the length of bench-terraces did not influence significantly the run-off, soil loss and the total yield of potato growth on them. However, from the analysis of the fortnightly soil-moisture data, it was observed that on benches longer than 91.5 m, the farther ends of the terraces were significantly drier than those near the cross disposal drains. On this consideration, it was recommended to have bench terraces, of about 100 m in length in the Nilgiris. Five longitudinal grades of bench terraces (*viz.* 0.15, 0.33, 0.50, 0.60 and 0.84 per cent) constructed on land having 25 per cent slope and with 2.5 per cent inward slope were compared to Ootacamund.

The run-off and the soil loss increased significantly with the increase in the grade. However, the maximum run-off was only 0.9 per cent of the total rainfall and the soil loss was only 0.3 tonne/ha under the maximum grade of 0.84 per cent, both of which are negligible.

The differences among the yields of the potato crop from bench-terraces of various grades were also not significant. Similarly, the effect of the inward grade of the bench terraces (*viz.* 1, 2, 4, 6, 8 and 10 per cent) on the yield of potato was not found to be significant.

GRASSED WATERWAYS

A grassed waterway is associated with channel terraces for the safe disposal of concentrated run-off, thereby protecting the land against rills and gullies. A waterway is constructed according to a proper design and a vegetative cover is established to protect the channel section against erosion because of the concentrated flow. In the alluvial soils of the Doon Valley under the humid subtropical climate, Panicum repens was the best-suited grass, followed by Brachiara mutica, Cynodon plectostachyus, Cynodon dactylon and Paspalum notatum. The suitability of a grass was based on the cover it gave, the ease with which it was established and the forgae yield obtained from it.

Evaluation of Effects of Soil and Water Conservation on Agricultural Land

The beneficial effects of conserving soils moisture by constructing bunds and levelling the land have been extensively demonstrated in the semi-arid alluvial plains of Uttar Pradesh where 35 per cent, 63 per cent and 98 per cent increase in the yields of kharif and rabi crops has been obtained by bunding alone, levelling alone and bunding-cum-lenelling respectively. Bunding has increased the yields of Setaria, cotton and johar by 18 per cent, 11 per cent and 17 per cent in large-scale field trials in Tamil Nadu.

Contour-bunding has not only increased the yield of crops in the dry farming areas in Maharashtra, but has also increased the water level in the wells and the employment oppurtunities as the age of bunding increased.

Estimation of Peak Rate of Run-off

For designing various structures and carrying out mechanical measures to control erosion for the safe disposal of run-off, such as grassed waterways and check dams, the estimation of peak rate of run-off is an important design criterion.

Three monographs have been developed to estimate the peak rate of run-off by using three different method\s, *viz.* The Rational Method, the Cooks Method and the Hydrologic Soils-Cover Complex Method.

Gully-Erosion Control and Reclamation

It is estimated that about 2.3 m ha (5.75 m acres) of land in India is affected by severe gully erosion.

These gullies are an indication of very bad management of the land resources. Apart from the fact that the gullied land has been completely destroyed, the gullies are a menace to the adjoining tablelands. Whereas, as a result of gully erosion, the land resource, one of the most important natural resources we have, is completely lost, the people along the river banks suffer from innumerable hardships.

The ravine lands pose a socio-economic problem as they harbour dacoits and undesirable social elements.

Also, the ravine lands are one of the chief sources of sediment which chokes the reservoirs in the country.

The gently sloping nature of the land, the loamy-sand, the sandy-loam and the silty-loam texture of the soil, intense rains in the monsoon, the improper land use by resorting to overgrazing, the biotic interference with the natural vegetative cover and the faulty agricultural practices are the chief causes of gully erosion all over India.

Classification of Gullies

Any system of gullies has an independent catchment, with a regular stream which has been termed the "drainage system". In each drainage system, it has been observed that gullies with defined side slopes, bed width and depths occur in a regular order. In the upper reaches of the drainage system, the gullies are wide and shallow, with varying side slopes.

The middle part of the drainage system is usually deeper, wider and has uniform side slopes normally up to about 15 per cent.

The lower portion of the drainage system which is nearer to the main river is usually very deep, has steep side slopes and is associated with intricate branch gullies. The gullies have been classified as follows:

Symbol	Description	Specifications
G_1	Very small gullies	Up to 3 m deep. Bed width not greater than 18 m. Side slopes vary.
G_2	Small gullies	Up to 3 m deep. Bed width greater than 18 m. Side slopes vary.
G_3	Medium gullies	Depth between 3 and 9 m . Bed width not less than 18 m. Side slopes uniformly sloping betweeen 8 and 15 per cent.
G_4	Deep and narrow gullies	(a)3 m to 9 m deep. Bed width less than 18 m. Side slopes vary. (b)Depth greater than 9 m. Bed width varies. Side slopes vary, mostly steep or even vertical; with intricate and active branch gullies.

While the general pattern of land-capability classification is based on the recommendations of the Soils-Survey Manual, the interpretation in relation to slope classes and the susceptibility to erosion hazard by active gullies has been modified to suit the conditions in ravine lands. In ravine lands, the slope of the land and the nearness of gullies are the most important and dominant factors which vary widely and influence land capability. The land-capability classification specifications for the ravine lands, as influenced by the slope of the land and the gully erosion hazard, are given above: Land-capability classes II, III, IV, VI and VII are to be annexed by symbol "e", indicating damage by the past erosion, the present hazard and the future susceptibility to gully erosion.

Land-form	Slope (%)	Distance from gully rim (m)	Land-capability class
Tablelands	0-1%	Beyond 60	I
Tablelands	1-3%	Beyond 60	II
Tablelands and wide humps in between gullies	0-3%	Between 6 and 60	III
Tablelands	3-5%	Beyond 6	III
Tablelands	5-10%	Beyond 6	III
Tablelands	10-15%	Beyond 6	IV
Marginal land between the gully rim and the tablelands	0-15%	Within 6 m of the gully rim	VI
Tablelands	15-25%	Within 6 m of the gully rim	VI
Very small, small and medium gully sides and beds (G_1, G_2 and G_3	--	--	VI
Tablelands	25%	--	VII
Deep and narrow ies	--	--	VII

Control of Gullies and their Reclamation for Various Land Uses

The practices described below are arranged in the order in which they should be adopted; the cost of their execution increases progressively.

Closure to Grazing and Other Biotic Interference

A study of the natural flora of the eroded areas has shown that only poor types of annual and unpalatable grasses are growing in place of the desirable climax associations which should normally exist in the prevailing soil-climatic environment. It is reported that at Vasad in Gujarat, as result of closure to grazing and other biotic interference:

- Aristida funiculata and Themada triandra were first replaced by Apluda mutica (a tall annual grass) and then by the perennials, such as Eremopogon foveolatus, Heteropogon contortus, Dicanthium annulatum and Cenchrus sp.,
- The loss of soil and water progressively decreased from the area as the natural vegetation improved,
- There was not only qualitative but quantitative increase in the yield of grasses.

Similar results have been reported on the clay loam and clay soils of the Chambal ravines where closure to grazing and biotic interference resulted in the qualitative and quantitative increase of useful species, such as Dicanthium annulatum, Cenchrus ciliaris and Apluda mutica. Under the upper Damodar catchment conditions, it has been observed that closure to grazing reduced the soil loss from 3.3 tonnes/ha (under overgrazing condition) to only 0.6 tonnes/ ha.

Checking the Growth of Gullies

Contour and Peripheral Bunding

After thc closure of the ravine lands, the immediate problem to be tackled is to stop the devastating rate of soil erosion and the growth of gullies and the conversion of the cultivated tablelands into wastelands for this purpose, it is essential to retain as much precipitation a possible on the land in the semi-arid and sub-humid areas and safely dispose of the excess run-off in humid areas. In the semi-arid and the sub-humid area of India, contour and peripheral bunds of various cross-sections have been found ideal for this purpose. In Gujarat, it is now recommended that for alluvial soils the contour and peripheral bunds should be 0.9 sq m to 1.2 sq m in cross section and should be spaced at 90 cm to 120 cm vertical intervals.

The contour and peripheral bunds are stabilized by sodding with Dicanthium annulatum, Cenchrus ciliaris and Panicum antidotale. It has been observed at Vasad in Gujarat that land directly under bunds can give a net annual income of ₹ 385 per ha from the sale of these grasses. The excess run-off is safely disposed of through grassed ramps or pipe outlets. In the Chambal ravines, where the soil are relatively fine and the rainfall is heavier, graded bunds in association with grassed waterways and drop structures for the safe disposal of water have been found to be better than the contour lands.

Gully Plugging

The eroding and deepening of gully beds can be prevented with gully plugs. Gully plugs protect the gully beds by reducing the speed of run-off water, redistributing it, increasing its percolation, encouraging silting and improving thesoil-moisture regime for establishing a plant cover.

Gully plugs of various materials, *e.g.*, brushwood live hedges, earth, sand bags, brick masonry and boulders have been tried in India. The size and material of the gully plugs depends on the width, length, and the bed slope of a gully and the anticipated run-off. All types of gully plugs are effective either in retaining or retarding the run-off. Earth is the cheapest and most readily available material and it is, therefore, easire and economical to construct the earthen gully plugs, wherever possible.

Boulder gully plugs are equally effective if the material is available. Brushwood is often available, but there is likely to be a shortage of wooden posts. Moreover, brushwood gully plugs cannot survive the severe attack of white ants, in spite of the fact that the wooden posts are coated in tar, they disintegrate in a couple of years. In the long run, they prove to be costly.

For the same considerations of short life and high costs, the sand-bag gully plugs are not recommended. Gully plugs of brick masonry are constructed at the confluence of all gully branches of a compound gully. For gullies where no run-off is expected from the top, earthen gully plugs of 1.1 sq m cross-section (with a grassed ramp, 22.5 cm below the level) and spaced at 45-60 m horizontal intervals are suitable; for gullies in which run-off from the top is expected, an earthen gully plug of 2.2 m cross-section with a pipe outlet is to be provided.

The diameter of the pipe shall be determined by the catchment area (15 cm diameter A.C.C. spun pipe up to a discharge of 2 to 3 cusecs from an effective catchment of 1.6 ha is suitable). A composite check dam of earth and brick masonry (spillway portion) is necessary for larger catchments (more than 1.6 ha). This is located at the confluence of a big compound gully with the main drainage system or in the bed of the main drainage system at 1.2 m vertical intervals or 120 m horizontal intervals.

Reclamation of Gullies for Cultivation

The belief that all the gullied land can be reclaimed for cultivation by making level benches in gully beds is not feasible.

This may be possible if no limit is fixed on expenditure. It has been observed that small and medium gullies can be conveniently, safely and economically reclaimed for cultivation, whereas deep and narrow gullies should be retired to a permanent vegetative cover of grasses and trees.

Reclamation of Small Gullies

Small gullies are reclaimed by clearing, minor levelling with bulldozers and constructing diversion-cum-check bunds of 1.5 sq m cross-section spaced at horizontal intervals of 30-45 m. Grassed ramps are provided for the disposal of excess run-off at the ends of bunds near the gully sides.

Consecutive grassed ramps are diagonally opposite to one another so that the path of run-off is lengthened and its flow velocity is reduced so as not to cause any scouring. At the end of the small gully, a composite check dam of earth and brick masonry spillway portions constructed.

Reclamation of Medium Gullies

A small gully gets transformed into a medium gully along the length of the main drainage system. A medium gully is reclaimed by clearing and levelling the bed and constructing a series of composite earth and brick masonry check dams at vertical intervals of 1.2 m (which gives a horizontal interval of 120 m on 1 per cent slope of the gully bed) and terracing the side slopes.

Design of Composite Check Dams

The design discharge (maximum run-off) through the drainage channel is computed from the empirical formula:

Q = 0.0028 C.I.A.

Where

Q = Run-off in cubic metres per second (m^3/sec)

I = Rainfall intensity in mm/hr

A = Watershed area in ha

The length of the spillway is determined from the formula:

Q = 1.70 LH3/2

Where

Q = Run-off in cubic metres per second (m^3/sec)

L = Length of the weir notch in metres H = Height in metres of water flow above the spillway level

Terracing of Side Slopes

The uneven side slopes of the medium gullies having 8 to 15 per cent slope are bench-terraced into level terraces at 0.9 to 1.2 m vertical intervals. The terraces are given a back slope of 1 in 50 and a longitudinal grade of 1 in 200 towards the grassed outlet. A ridge bund of 0.3 sq m cross-section is provided at the edge of each terrace. Terrace faces are given a slope of 1.5: 1. Bench terraces are constructed when the gully sides are having a uniform slope for a length of at least 120 m to justify the cost of terracing. The terrace faces, grassed outlets and earthen checked dams are stabilized by sodding or growing them with suitable grasses. Dicanthium annulatum and Cenchrus ciliaris have been found suitable for this purpose, specially in Gujarat and in the ravine son the banks of the Jamuna river. Bench terraces and the check dams require careful maintenance for the first two years in view of the unsettled conditions of the soil.

Reclamation of Deep and Narrow Gullies

As mentioned earlier, the best land use of ravine lands is to retire them to permanent vegetation comprising grasses and trees. This type of land use is a must for the deep and narrow gullies. The natural tree species of the ravine lands of Gujarat comprise Acacia nilotica, A. senegal, A. leucophloea, Azadirachta indica, Albizia lebbeck, Feronia elephantum, Prosopis spicigera, etc. With closure to grazing and other biotic interferences, these species make very good growth which is further helped by silvicultural operations, such as climber-cutting and thinning.

Afforestation with a number of species has been tried. Acacia nilotica, Ailanthus excelsa, Albizia lebbeck, Azadirachta indica, Dalbergia sissoo, Dendrocalamus strictus, Eucalyptus camaldulensis, E. citridora, E. hybrid, Pongamia glabra, Phyllanthus embilica, Salmalia malabarica and Tectona grandis are quite promising species for ravines in Gujarat.

For the afforestation of ravines along the banks of the Yamuna river at Agra, Dalbergia sissoo, Acacia nilotica, Azadirachta indica, Albizia lebbeck, Pongamia pinnata, Holoptelia integrifolia, Dendrocalamus strictus and Eucalyptus hybrid have been found to be the most suitable species.

In Kota, Rajasthan, Dendrocalamus strictus, Salix tetrasperma in gully beds and Acacia nilotica on the top and marginal lands have been found to be very suitable.

Landslide Control

Landslides pose an immense threat to highways, villages, agricultural lands, lines of communication, etc. The problem of landslides is more serious in the Himalayan region than anywhere else in India. One landslide about 4 ha in area in Nalotakhala watershed at Dehra Dun has been controlled.

The Nalota Nala landslide was controlled by protecting the landslide against the scouring action of the torrent (by training the lower reaches of the torrent, stabilizing the debris cone, stabilizing the middle reaches of the torrent), by stabilizing the landslips and landslides (by wattling to provide temporary support to bare erodible slopes with grass and brush, by planting deep-rooted shrubs and grasses, such as Pennisetum purpureum, Ipomoea carnea, Vitex negundo and Pueraria hirsute).

Based on the observations on the landslide control, it is recommended that the control methods may be adopted, depending on the kinds of materials available in the area, the cause of the problem, the type and the rate of the movement of the slide. These factors vary so greatly from slide to slide that details of remedial measures are required to be tailored to deal with each slide.

Torrent-training and Stream-bank Erosion

The torrent problem is of immense magnitude, extending all along the foothills and the plains abutting the Himalayas and the Siwaliks in Northern India. A study was, therefore, undertaken at the Soil Conservation Research Centre, Dehra Dun, in the lower reaches of a torrent (Bainkhala Torrent) to develop a co-ordinated technique based on suitable vegetative and engineering measures for torrent-training and stream-bank protection. The Bainkhala Torrent-treatment measures were observed for 10 years. Their performance and utility were evaluated as under:

- Out of the various species tried, for planting live hedges and rainforcing spurs, Vitex negundo was found to be most suitable followed by Arundo donax and Ipomoea carnea. Ipomoea carnea in the case of its planting along spurs was found to be damaged by the abrasive action of debris and boulders moving with flood waters.
- The species found suitable for planting on sloped banks were Aristida cynantha, Chrysopogon fulvus, Arundo donax, Pennisetum purpureum (Napier) and Cynodon plectostachyus (giant star grass).
- Boulder rip-raps on sloped banks invariably gave way owing to scouring at the toe, in spite of the 1 m deep foundation. Even the planting of a live hedge at its base was not much help, unless the main current was deflected from the bank by erecting spurs, etc.
- Jetted posts and jacks provided good protection against bank erosion by inducing siltation, but these structures were usually damaged by the villagers and graziers who often took away the wooden posts for use as fuel. Jacks were found particularly suitable for a straight reaches of the banks.
- The repelling type of spurs were found to be most suitable for quick reclamation of concave banks, since, these spurs induced heavy siltation both on the upstream and downstream sides. However, they needed special protection of their nose which was subjected to excessive scouring.
- The wooden posts in the log spurs and cross barriers started rotting after 2 to 3 years.
- Out of the various species tried for wooden posts, Salmalia malabarica, Cordia myxa and Broussonetia papyrifera gave better performances. It was seen that shoot development took place in most of the species tried, but it was not followed by root development. The shoot died within a few months. On the whole, posts from younger branches were better in sprouting and surviving than those from the older branches.
- Out of the various types of spurs tried, the boulder spurs reinforced with vegetation and gabion spurs (both with 'nose' protection) gave better performance.
- The best months for erecting vegetable structures and making planting was found to be in January and February, when some winter showers are received and the vegetation can get sufficient time to get established before the floods during the monsoon.

- Pennisetum purpureum (Napier) was found to be the best species for stabilizing fresh deposits of silt near the spur and other structures. For a site having coarse deposits (shingle and boulders), Aristida cynantha and Saccharum spontaneum proved to be useful. The data on the forage yield of Napier from the torrent bed have shown that an average yield of about 4,000 kg/ha can be obtained.
- For the utilization and stabilization of the reclaimed areas behind the banks, fuel-cum-fodder plantations of Dalbergia sissoo (sisham) and Chrysopogon fulvus (gorda grass) were found to be quite suitable.

TILLAGE SYSTEMS FOR SOIL AND WATER CONSERVATION

The care of soil is essential for the survival of the human race. It provides most of the food required, fibres for clothing and wood for building materials. But in many parts of the world the soil has been so damaged by mismanagement that it will never be able to produce, at least not in the near future. This maltreatment has led to soil erosion on a vast scale, that is, physical loss of upper layer of soil which is most vital for plant nutrition.

Severe soil erosion has taken place even in advanced countries. It has become a global problem and its remedy lies in the application of soil and water conservation practices through a soil conservation programme. However, the farmers themselves have to carry out the necessary practices on their land as a part of their farming system.

In the early stages of the development of crop culture, vegetation that surrounded and competed with plants (that provide food) was flattened and removed by hands.

Food production was generally limited under these conditions. As man learned to grow certain food plants in the more desirable locations, it became necessary to remove existing vegetation which heralded the beginning of soil tillage.

For tilling the soil in older days, crop producers used crude implements made of wood or stone. With the help of the implements, they used to loosen the soil but gradually they realized that weeds are primarily responsible for restricted growth of plants.

Thus, improved implements and techniques for weed control have been developed. Weed control is one of the basic reasons usually given for tilling a soil.

In the beginning the agricultural practices were simpler and the tillage operations were also simple. But in the modem agriculture, tillage is defined and described as follows.

The operation of implements through the soil to prepare seedbeds and root beds, control weeds, aerate soil and cause faster break- down of organic matter and mineral to release plant nutrients.

TYPES OF TILLAGE

Different types of tillage systems are given here:

- *Clean tillage:* Cultivation of a field so as to bury all plant residues and to prevent growth of all vegetation except that of the desired crop.
- *Conservation tillage:* Any tillage sequence that reduces soil or water loss relative to conventional tillage. It is often a form of non-inversion tillage that retains protective amounts of crop residues on the surface.
- *Conventional tillage:* The combined primary and secondary tillage operations normally performed in preparing a seedbed for a given crop in a given geographical area.
- *Minimum tillage:* The minimum soil manipulation necessary for crop production or meeting tillage requirements under existing soil and climatic conditions,
- *No tillage:* A method of planting crops that involves no seedbed preparation other than opening the soil for the purpose of placing seed at tile intended depth; usually involves opening a small slit or punching a hole in soil; usually involves no cultivation during the growing season; usually involves chemical use for weed control, Also called as slot planting, zero tillage, direct drilling,
- *Reduced tillage:* A system in which the primary tillage operation is performed in conjunction with special planting procedures to reduce or eliminate secondary tillage operations; less than conventional tillage. Similar to minimum tillage, sometimes called limited tillage.
- *Stubble mulch tillage:* A system of tillage that retains the stubble of crops or crop residues in place on the land, thus providing a protective surface cover before and during seedbed preparation and at least partially during the growing season of the succeeding crop.

OBJECTIVES OF TILLAGE

Although soils are tilled for controlling weeds and also for other reasons, the very goal of tillage is to stabilize and increase crop production. Tillage is the physical manipulation of soil with tools and implements to result in good tilth for better germination of seeds and subsequent growth of crops. In other words, it is the tilling of land which is done for bringing about conditions that are favourable for the cultivation of crops. Both implies to the physical condition of the soil in relation to plant growth. Tilth is brought by tillage.

Tillage has got several objectives:

- The main objective of tillage is the preparation of land to make it suitable for crop cultivation. It makes the soil loose and friable with sufficient aeration and moisture for the germination of seed.
- Tillage helps 10 control weeds and incorporate the weeds, crop residues, green manures and other organic manures and fertilizers,

soil amendments and other agro chemicals applied for the control of weeds, insect pests etc.

- It improves soil structure, soil aeration, water holding capacity of soil, soil permeability, organic matter decomposition and creates suitable conditions for root penetration removing hard pans, if any, in the soil.
- The most important objective is to prepare good seed bed.
- Proper tillage results in soil and moisture conservation through high filtration, reduced run-off and increased depth of soil for moisture storage.

To emphasize upon the soil and water conservation, the need of the hour, one has to adapt tillage systems that are most suitable for the prevailing soil, crop and climatic conditions; Under some conditions a simple change in tillage method may be adequate.

SELECTION OF TILLAGE SYSTEMS

The tillage system selected for a particular situation depends on climatic zone, crop to be grown, soil factors, economic level of the producer, preference of the producer etc.

Climatic Factors

The climatic factors like rainfall, temperature, radiation and wind are the most important ones while considering the tillage system to be adopted.

Rainfall

In areas where rainfall is low, tillage systems and related practices for the conservation of water are highly desirable because as much moisture as possible must be stored in soil for the use of plants for their growth. In order to achieve this, tillage systems which can improve the water infiltration rate, can trap moisture from snow and dew and can reduce the evaporation from the soil surface should be adopted.

Infiltration can be enhanced by reducing run-off rates, making the soil surface condition suitable for rapid infiltration and removing soil profile layers which restrict the water movement down into the soil. Reduced evaporation can be achieved by improving the micro climate at the soil air inter-phase thus increasing the storage of water in the soil. Mulching not only helps in improving the microclimate but also intercepts the radiation, avoid high soil temperature and make the soil rough which in turn reduce the wind speeds.

In scanty rainfall areas, water and wind erosion are the major problems which can be controlled by the practices mentioned above.

In regions where rainfall is high, use of grassed waterways is desirable as they conduct excess water away from the fields.

Temperature

The prevailing temperatures strongly influence crop adaptation in a region. Tillage systems have little or no effect on air temperatures, but can be used to manipulate the soil. This should be particularly useful in areas where frost-free periods are relatively short and need few more extra days of growth period. Mulching will help in maintaining the soil temperatures.

A surface mulch decreases day time soil temperature but a combination of mulching and ridging in the field gives the effect of a tillage and can maintain soil temperature and soil water content and also improves the use of conserved water, thus protects soil from erosion.

Radiation

Tillage affects the radiation balance of a soil mainly through its effect on soil colour and water content, slope relative to the sun and surface residues. Light coloured surfaces reflect radiation whereas dark ones absorb it. Thus, a freshly tilled, rough moist soil will absorb more radiation than a smooth, dry soil. On sloping soils, absorption is greatest when the sun is perpendicular to the slope.

The radiation balance due to surface residues is affected by the colour, decomposed stage and geometry of residues. Bright residues reflect a large percentage of the incoming radiation. As residues decompose and darken, reflectance decreases. The radiation balance has a direct effect on photosynthesis. It also indirectly affects water conservation through its effect on soil temperature.

Wind

Wind speed has a major influence on soil. It is advised to take up cultivation or tillage systems suitable for a particular climatic zone mainly because of the potential of different tillage systems for controlling wind erosion.

On soils subject to wind erosion, tillage systems that maintain surface residues or provide a rough, cloddy or ridged surface during the windy season should be used. Where winds from one direction predominate, the direction of tillage operations that ridge the soil surface should be perpendicular to the prevailing winds. On soils where the wind erosion potential is much greater than water erosion potential, soil ridging tillage should be perpendicular to the prevailing winds than on the contour. While contour tillage would control water erosion, wind blowing parallel to the ridges could cause greater erosion than that caused by water during infrequent rainstorms. This would be especially true on deep sandy soils having relatively high infiltration rates.

Crops to be Grown

Different tillage systems are adopted for different crops. Differential crops responses are mainly due to the tillage systems adopted. They affect soil water content, weed control, aeration, root zone depth, fertility etc. Some crops like fibrous rooted grain crops are not affected by the soil physical conditions resulting from tillage methods but their growth is related to the amount of water available in the soil. In contrast, root crops such as sugar beet, responses to tillage induced changes such as water infiltration rate, aeration, bulk density changes etc.

Tillage systems that provide a specific soil condition may also be required for some other crops *i.e.* for tuber and root crops a deep loose rootzone is required and small seeded crops requires uniform finely granulated seed bed. Apart from these, short statured crops like vegetables and cotton, trash free surface is essential as they may interfere with harvesting operations or lower crop quality.

Soil Factors

Soil factors like slope, texture, depth, density, salt or alkali content and drainage are important in selecting the tillage systems.

Slope

On nearly level or gently sloping soils, many tillage systems provide the desired conditions and effectively conserve the soil and water resources. As slope increases the choices of tillage systems become more limited. A conservation tillage system that. maintains surface residues is best suited for water erosion control on sloping soils when other supporting practices are not used. When supporting practices such as contouring, terracing and strip cropping are used, tillage is possible on soils of somewhat greater latitudes. Land leveling permits the use of almost any type of tillage systems. However, on steeply sloping soils, the leveled areas become extremely narrow and therefore the use of some tillage methods or systems may be impractical. Tillage may even be restricted to use of hand implements on such areas.

Texture

Soil texture has a major influence on a soil's susceptibility to erosion and therefore, on the selection of tillage methods or systems for controlling erosion. Conservation tillage systems, as a rule, are highly effective for controlling wind and water erosion on all types of surface textures.

Sandy soils, in which very less crop residues are available, are susceptible to wind erosion. Normally chisel plough is used on these soils than mould board or disc plough. To be effective the tillage should be performed perpendicular to the prevailing winds and at a soil-water Content conducive for forming a rough surface. Clods should not be pulverized. Mould board ploughing is sometimes used on sandy soils to bring finer textured material to the surface to aid in forming clods and, thereby reduce susceptibility for erosion.

The choice of tillage method or system is greater on finer textured soils for controlling wind erosion. The major pre-requisite of any tillage system for controlling wind erosion where surface residues are not available. is to produce a rough, cloddy or ridged surface. About 75 per cent of the surface soil material should be greater than 0.84 mm in diameter to reduce the soil losses due to wind erosion.

Effective control of water erosion usually depends on the use of supporting practices such as contouring, terracing, strip cropping and crop rotations in conjunction with the tillage system. Contouring and terracing reduce slope gradient and length and reduces the loss by run-off. Strip cropping, depending on the crop in the strip reduces run-off velocity and traps the sediments. The land area, where crop rotation has been practiced constantly, would become somewhat resistant to erosion because, there would be less soil dispersion, less surface sealing and greater water infiltration and therefore, less run off and erosion.

Sometimes the tillage system itself may decrease the soil residues. In general, best control is obtained with tillage that maintains unsealed soil surface and permits high water infiltration rates. For this purpose presence of water-stable aggregates are desirable which normally result from maintaining, as much organic material as possible at or on the soil surface. When precipitation rates exceed infiltration rates, temporary surface storage of water reduces run-off and aids in erosion control.

Ridge forming tillage on the contour is a long proven run-off control and water conservation practice. Run-off can be eliminated on the contour of gently sloping land by adopting lister ploughing (lister plough is a plough with double mould board) in conjunction with closed-end level terraces or by basin listing. For subsequent temporary storage on the surface, water collected behind the ridges or dams must infiltrate before the next precipitation (rainfall) event.

In sandy soils, by increasing the clay content water holding capacity can be increased. If too much clay is placed in the surface layers, it may decrease the infiltration rate. However, at least 8 per cent clay is needed in the surface layer of sandy soils for subsequent tillage to result in clods that resist wind erosion.

Depth

Selection of tillage systems based on soil depth is mainly concerned with the depth to an untillable layer (like bed rock) or depth to a layer that would contribute undesirable sub- stances like sand, gravel, rocks, high calcium materials, saline or alkali materials or strongly acid materials etc., to the tillage zone.

On shallow soils, non-inverting tillage should be used to minimize damage to equipment or the danger of contaminating surface soil with the undesirable substances. Rolling type equipment such as disc harrows, ploughs, disc openers planters are generally well adapted for use on shallow soils underlain by rocks. However, disc tillage is not desirable because it accelerates erosion. Under shallow soil conditions, a conservation tillage system involving minimum soil distrurbance, such as use of disc opener planter and applying herbicides for weed control would be most desirable.

Any tillage system that provides desired soil condition and conserves soil and water resources can be used on deep soils.

Dense Horizons in the Soil Profile

If there are any compact layers in the profile they affect the water infiltration rate. These compact, restricting layers like hard pan, fragipan, plough sole etc. may be formed from tractor and implement operations or animal trampling or can be formed naturally also.

Hard pans having rock layers, generally are not disrupted. However, fragipans, plough soles etc. can be broken by chiselling or by doing deeper than normal ploughing. Response to those operations would be greater when performed with the soil relatively dry. Dense or compact zones when disrupted increase water infiltration, conservation, greater root penetration and proliferation to extract water from the soil.

Alkali or Salt Content

Alkali soils can be reclaimed by adding gypsum to the soil which usually gives the greatest benefit with respect to water infiltration, because harmful salts would be leached down.

Apart from all these factors economic level of farmer to adopt tillage systems and his reference should also be considered.

TILLAGE SYSTEMS

Since man began to cultivate the soil many tillage systems have been come into practice of which important ones are described here.

Shifting Cultivation

This is a widely practiced system in different parts of the country. Major advantages of this system are the low capital input requirement, dependence on natural or regenerated soil fertility and the opportunity to grow a wide variety of crops.

As it involves clearing of the forest land by fire, it destroys weed seeds and to some extent weed control is being achieved. But non- application of fertilizers and herbicides results in declining soil fertility and greater weed problems. Also burning of forest litter, crop residues, grasses etc. may control the weeds but make the soil susceptible to erosion. As the fertility declines, there is a need to change the plot in every 2 to 5 years.

Labour Intensive Continuous Cultivation

Continuous cultivation refers to annual cropping where one or two years of occassional fallowing may be there unlike shifting cultivation in which land is fallowed for a longer period than the cropping period.

There are two systems that comes under this cultivation:

- Subsistence continuous cultivation: Labour intensive continuous cultivation at the subsistence level is similar to shifting cultivation in most regards except that a plot is cropped for a longer period than it is fallowed. Farmers mainly produce sufficient quantity for their family. Because a plot is cropped continuously major clearing is not necessary in each season. However, greater. effort is usually required to control weeds because weed problems are generally increased with cropping period.

  The basic tools used in this system are the hoe and machete (a large heavy knife) or cutlass (a short, heavy slightly curved knife).

  Under poorly managed conditions with no provisions for soil fertility maintenance, and soil and water conservation, crop yields decrease and land degradation intensifies. With improved management, such as soil fertility maintenance through recycling of various organic wastes, growing soil improving crops in rotation, use of other resource conserving practices etc. good crop yields can be obtained without serious land degradation.

  In order to maintain the productivity, fertilizer application, use of mulches, adopting soil and water conservation practices, can be put into practice.

- Extra production continuous cultivation: This system is also based on hand labour for the necessary cultural operations. The basic tools are spades, hoes and also some other implements are used.

  Under this system farmers normally produce more than they require for their families. They also grow one or two crops at commercial level. Animal production may also be involved in association with crop production.

  Advantages of extra production continuous cultivation are; conservation of soil and water resources, improves soil fertility, recycling of organic wates, and improve soil physical condition. Recycling of organic wastes is labour intensive. Mixing residues with soil by use of hand implements is also difficult. Therefore, the development of no tillage systems for arable crop production is practiced for efficient recycling of organic wastes. Under no tillage system of cultivation, soil loss is less as it involves minimum disturbance to the soil. Maintenance of organic matter is also undertaken as it increases the water holding capacity, infiltration into the soil.

  Major disadvantages are high labour requirements, limited arable land for expansion, limited organic wastes to manage, costly or limited fertilizers etc.

In labour intensive continuous cultivation systems, hoe cutlass or machette and spade are the main implements used.

Hoes are the major weed control implements and hoeing helps in smoothening the surface, breaking and pulverizing the clods and spades can also be used in loosening and mixing the soil.

Animal-drought and Small Tractor Cultivation

The era of animal draught technology for agriculture began about 3000 BC when man learned to harness the animals he had domesticated earlier. The wheel was discovered which eventually lead to the development of carts pulled by humans and animals and ultimately tractors were developed for tillage operations. Essentially any soil condition achieved by use of animal or tractor power can be achieved by hand labour. But the major difference lies in the labour required, timeliness of the operations and intensity with which soil is tilled.

Once the soil is tilled for controlling the weeds the moisture may get evaporated from the soil. As for weed control, timeliness of seed bed preparation also can have a major effect on water conservation. Seed beds should be ready whenever conditions are optimum for establishment of the given crop. This can be achieved by performing major tillage operations for seed bed preparation well before the optimum planting date. This may require that the tillage be performed under relatively dry soil conditions, especially in regions having distinct low and high rainfall,regions. At some locations, light showers during the dry season are beneficial for seed bed preparation before onset of the rainy season. Final seed bed preparation can then be achieved without major delay at the start of the rainy season, thus resulting in more timely crop establishment and potential for using growing season rainfall more efficiently for crop production. Such early seed bed preparation and crop establishment is conducive to water conservation in several ways. So, early tillage helps in water conservation.

In this system, the need for capital increases. When the cultivation system is subsistence or labour intensive cultivation system, the need for capital is low, because most of the labour is supplied by the farmer or family members and the implements used are simple. The need for capital increases with the introduction of animals or small tractors into the cultivation systems. Capital is required by the farmer to acquire the animals or tractor, satisfactory implements, to provide food for the animals and fuel for the tractor and to cover other production expenses.

Modem High Technology Cultivation

Modem high technology refer to those systems in which most crop production operations are performed with fuel powered machines like tractors (include harvesters, balers etc) and other implements and machines, with the aid of chemicals (herbicides, insecticides etc.) Labour is limited mainly because many operations are carried away by the machines and manual labour may be necessary for some crops to reduce plant population and for weeding and harvesting. In most of the cases, crops are planted at a proper seed rate to give desired plant population and weeds are controlled by applying herbicides. For harvesting of crops also mechanical means are used but hand labour is still widely used and even necessary for harvesting some crops.

A major advantage of this system is that field operations can be accomplished rapidly when conditions become favourable. For example, family operations like ploughing, crop establishment and harvesting can be completed quickly when conditions are optimum. But energy and equipment requirements are high in this system. Chemicals play an important role and are widely used to control weeds, diseases and insects. Under favourable plant and climatic conditions, one or a few correctly made applications of herbicides can effectively control weeds that would normally require one or more tillage operations and in many cases, many additional hours of hand labour. For example, in wheat-fallow-sorghum cropping system, three to four tillage operations are required for weed control.

PLAN REQUIREMENTS FOR AGRICULTURAL ACTIVITIES

They may keep their soil in place by preventing accelerated erosion or they may capture soils that have eroded at an accelerated rate to prevent them from polluting the waters of the Commonwealth. To keep soil in place, one option is to develop, implement and maintain a *conservation plan* for the lands on their crop producing fields. To capture soils that have eroded at an accelerated rate, farmers may design, implement and maintain an *erosion and sedimentation control plan* consisting of appropriate numbers and locations of sediment removal best management practices. Such a plan, while meeting the requirements of Chapter 102, does nothing to maintain the soil resource base that is required for a successful farm to remain successful over time. The requirements for each of these plans is detailed below and in PART III of this manual.

New building construction, road construction, etc. requires an erosion and sedimentation control plan to be prepared and implemented. Typical elements of an erosion and sedimentation control plan for these types of activities may include temporary diversions to channel upslope run-off around the disturbed site, and sediment removal devices (silt fence, sediment trap or basin) to treat the sediment laden run-off downslope of the site

A *conservation plan* or an *erosion and sedimentation control plan* is prepared and implemented for all operations considered to be agricultural activities that involve earth disturbance and woodlots (during timber harvesting operations) on a farm. The plan should be available at all times and updated as needed to guide the responsible person or persons for the overall operations of the farm. Elements of the *conservation plan* include appropriate management and/or structural best management practices mentioned in Part IV of this manual.

Installation of conservation practices which disturb the land surface, such as diversions, waterways, etc. is also required to have erosion and sedimentation control measures designed and installed as part of the conservation practice installation. Details and descriptions for erosion and sedimentation control measures are found in the Erosion and Sediment Pollution Control Programme Manual.

A *conservation plan* or an *erosion and sedimentation control plan* reflects the current agricultural operation. It is the responsibility of the landowner or user to keep the plan current. For example, if new fields are brought into production, the *conservation plan* or an *erosion and sedimentation control plan* is revised to include the new fields. Similarly if fields that were formerly permanent hay are converted to row crops, the *conservation plan* or an*erosion and sedimentation control plan* for those fields will need to be revised. Revisions to the *conservation plan* or *erosion and sedimentation control plan* must then be implemented. *How can you tell if you need to revise your plan, or if your plan is working?* Examine the condition of all waterways. Examine the condition of any watercourse flowing through or adjacent to your property. Sediment accumulation in the form of deltas at stormwater discharge points, sediment covering the stream bottom and aquatic vegetation may indicate accelerated erosion and resulting sedimentation from your property. Observe your crop fields during a rainstorm. Gullies forming and soil washing away from your crop fields are indicators that your *conservation plan* may not be adequate. Very muddy water running into watercourses adjacent to your crop fields are also an indicator of an inadequate *conservation plan*. Most practices included in a *conservation plan* are designed for a 10 to 25 year frequency storm. Some damage to practices can be expected from more intense storms; likewise muddy water discharges may occur from these storms even though an adequate conservation plan has been developed and implemented.

What do you do now? A farmer may receive assistance in preparing or revising a *conservation plan* by becoming a conservation district cooperator. The conservation district requests a representative of the Natural Resources Conservation Service, USDA, to provide technical assistance in preparing a *conservation plan* to meet the needs of the landowner and the standards of the Pennsylvania Soil and Water Conservation Technical Guide. Private consultants, Conservation District staff, and Certified Crop Advisers may also assist farmers to develop *conservation plans*. Call your local conservation district for further instructions.

The plan covers all agricultural activities involving earth disturbance and is approved by the conservation district. The landowner is responsible for the implementation of the plan. *Conservation plans* developed by landowners and consultants shall be submitted to the conservation district for review and approval and maintained for reference purposes in the conservation district files.

AGRICULTURAL EROSION AND SEDIMENTATION CONTROL PLAN CONTENT

Conservation Plan Content

A *conservation plan* shall be prepared by a person trained and experienced in erosion and sedimentation control techniques to meet the requirements of the Chapter 102 regulations, and includes the following:

- Sufficient management and/or structural best management practices to minimize accelerated erosion, prevent sediment pollution to the waters of the Commonwealth, and maintain the resource base. Generally this will require a*conservation plan* that meets the soil loss tolerance, commonly called "T". Soil loss tolerances denote the maximum level of soil erosion that allows high levels of sustainable economic crop productivity. The Revised Universal Soil Loss Equation (RUSLE) is a management tool used to estimate soil losses by sheet and rill erosion from rainfall. The purpose of the RUSLE is to predict long-term average soil losses in run-off from specific field areas under specified cropping and management systems. The effectiveness of a particular land treatment alternative can be evaluated when the predicted soil loss for that treatment is compared with the "T" for the specific soils involved. Additional information about "T", and RUSLE can be found in the *Pennsylvania Soil and Water Conservation Technical Guide*
- Computations to show that the flow velocity for storm water run-off from the farm for a 10-year frequency storm (or a discharge of 2.75 cubic feet per second per acre drained) will not cause erosion in on-farm or off-farm waterways or channels. See Chapter 4 of *Pennsylvania Erosion and Sediment Pollution Control Programme Manual* or Chapter 2 of the *NRCS Engineering Field Handbook* for additional information about these computations.

Agricultural Erosion and Sedimentation Control Plan Content

An agricultural *erosion and sedimentation control plan* should include best management practices to remove sediment from run-off where run-off from tilled areas flows directly into perennial streams. The requirements for best management practices are found in the *Pennsylvania Soil and Water Conservation Technical Guide* and in Chapter 4 of *Pennsylvania Erosion and Sediment Pollution Control Programme Manual.*

Plan Elements

A *conservation plan* consists of a conservation plan map, a soils map, and a narrative. An *erosion and sedimentation control plan* also contains these same elements. A sample *conservation plan* is included in the Appendices.

Conservation Plan Map - This map shows the farm boundaries, field boundaries, use of each field, acres in each field, permanent conservation measures installed and/or planned, and the scale of the map. Usually a scale between 1"=400' and 1"=1320' is sufficient as long as the intensity of the farm operation can be clearly shown. A title block on the conservation plan map includes the name of the farmer, the location of the farm (municipality and county), the acres of the farm, the date the map was prepared, the name of the person who prepared the map, and the scale of the map.

An aerial photograph may be used as a map. The map includes a legend that explains all of the symbols used. Soils Map - The soils map indicates the kinds and boundaries of soils. A title block on the soils map includes the name of the farmer, the location of the farm (municipality and county), the source of the soil survey data, the acres of the farm, and the scale of the map. A soil legend accompanying the soils map provides an explanation of all soil symbols shown on the map. A description of the mapping units includes the soil parent material, slope, texture, drainage characteristics, available moisture, depth, erosion hazard, and fertility characteristic

Conservation Plan Narrative - The narrative includes names, units and amounts, and locations of each of the conservation practices installed or to be installed. Implementation of the *conservation plan* means that the various measures be installed in a timely fashion. Certain measures are installed before others. For example a waterway may serve as an outlet for diversions. Therefore, the waterway is installed, stabilized and operating before the diversions are constructed. The year of installation for practices already installed; and a schedule (month and year) for practices not yet installed is included in the plan.

Continual management and maintenance of best management practices is essential for proper functioning. Structural practices are designed for a specific storm frequency and may suffer damage when larger storms occur. After such storms, practices are inspected and damage is repaired so they continue to function as designed and installed. Structural practices which require earth moving and stabilization should be frequently checked to assure that vegetation and other types of stabilization are adequate to protect the structure and to assure that it will continue to function properly. Reseeding and/or reshaping may be needed periodically, especially after major rainfall events. Vegetated practices should also receive adequate lime and fertilizer to assure that the vegetative cover is adequate for the intended purpose. After major rainfall events, they should be inspected for any type of damage or deterioration. Trees and brush should not be allowed to grow on berms, dams, or other structural embankments. Other best management practices such as stripcropping, contour farming and grass buffers require attention annually to assure the boundaries are maintained and adequate vegetation is maintained. Practices which are performed annually such as cover crops, conservation tillage, crop residue management and crop rotations, must be carried out according to the *conservation plan.* These practices require detailed attention to assure they are carried out when and to the extent specified. Tillage and residue management are two specific examples that require this attention.

Other Information to be Added to the *Conservation Plan* - Prior to the installation of structural conservation practices, detailed designs and construction drawings for the practices are prepared. Details of interim erosion control measures used during the installation of structural measures are included in the construction drawings.

An *erosion and sedimentation control plan* contains the same elements as a *conservation plan*. The Plan map would be similar to the one described for a *conservation plan*, except that the only practices to be shown would be the sediment removal best management practices. All other elements of the map would be as described above for a conservation plan map. The Soils Map would be identical to the one described above for a *conservation plan*. The Narrative would contain the details required for the construction of best management practices for sediment removal. It would also explain how the sediment removal best management practices would be maintained; where and how the captured sediment would be disposed of, and how it would be stabilized to prevent it's migration into the waters of the Commonwealth. Other Information would be similar to that described above for a*conservation plan*.

Management Practices Applicable to Agricultural Activities

The following list of conservation practices are some of the more common best management practices used to control erosion and sedimentation on farmland. Prevention of erosion is more desirable and effective than treatment of run-off which contains heavy concentrations of sediment. Diversion of surface water away from plowed and tilled areas and maintenance of cover on cropland through use of conservation tillage systems can be effective ways of preventing erosion.

Access Road - A travelway constructed as part of a conservation plan. It's purpose is to provide a fixed route for travel for moving livestock, produce, equipment, and supplies. It may also be used to provide access for proper operation, maintenance and management of conservation enterprises and conservation farming systems while controlling run-off to prevent erosion and maintain or improve water quality.

Barnyard Run-off Control - The collection and reduction of run-off water and agricultural wastes from barnyards, feedlots and other outdoor livestock concentration areas for storage or treatment to improve water quality.

Channel Vegetation - Establishing and maintaining adequate plants on channel banks, beds, berms, spoil and associated areas. This is done to stabilize channel banks and adjacent areas for temporary and/or permanent protection and reduce erosion and sedimentation.

Conservation Cover - Establishing and maintaining perennial vegetative cover to protect soil and water resources on land retired from agricultural production. This will help reduce soil erosion and sedimentation, thus protecting water quality and creating or enhancing wildlife habitat.

Conservation Cropping Sequence - An adapted sequence of crops designed to provide adequate organic residue for maintenance or improvement of soil tilth. By utilizing this practice, one will help improve the physical, chemical and biological soil conditions, maintain or improve soil productivity, protect the soil against erosion and overload run-off, and maintain or improve water quality.

Conservation Tillage System - Any tillage and planting system in which at least 30 per cent of the soil surface is covered by plant residue after planting to reduce soil erosion by water during the critical erosion period.

Contour Farming - Farming sloping land in such a way that preparing land, planting, and cultivating are done on the contour (this includes following established grades of terraces or diversions). This practice may be applied to reduce sheet and rill erosion, to manage run-off to increase plant available moisture and to improve surface water quality by reducing siltation.

Cover and Green Manure Crop - A crop of close-growing grasses, legumes or small grain grown primarily for seasonal protection and soil improvement. It usually is grown for one year or less to control erosion during periods when major crops do not furnish adequate cover.

Critical Area Planting - Planting vegetation, such as trees, shrubs, vines, grasses or legumes on highly erodible or critically eroding areas (does not include tree planting mainly for wood products).

Crop Residue Use - Using plant residues to protect cultivated fields during critical erosion periods. This practice maybe planned to reduce soil loss during critical periods, to conserve soil moisture, to maintain or improve soil tilth, to reduce fluctuations in soil temperature, and to provide food and cover for wildlife.

Cross Slope Farming - Farming sloping land in such a way that tillage, planting and harvesting are done perpendicular to the predominant slope, but not necessarily on the contour. This practice is used to reduce sheet and rill erosion and improve surface water quality by reducing siltation.

Diversion - A channel constructed across the slope with a supporting ridge on the lower side. This is done to divert excess water from one area for use or safe disposal in other areas. Establishment and maintenance of permanent stabilization (usually grass) is required.

Field Border - A band of grass or legume at the edge of a field used in place of end rows. This practice is used to control erosion and to protect edges of fields that are used as turnrows or travel lanes.

Filter Strip - A strip or area of vegetation for removing sediment, organic matter and other pollutants from run-off. This is done via several different processes such as filtration, infiltration, absorption, adsorption, decomposition and volatilization.

Forest Land Erosion Control System - Application of one or more erosion control measures on forest land. Erosion control system includes the use of erosion control structures, conservation plants and cultural practices for the control of sheet and rill erosion, gully formation and mass soil movement

Grade Stabilization Structure - A structure to control the grade and head cutting in natural or artificial channels. Waterways need stabilizing structures if slopes are so steep that the velocity of run-off exceeds the limit of protection that vegetation alone gives. Energy dissipaters may be required at the outlets of the structures. The structures should be designed and constructed to provide permanent stabilization. Types of grade stabilization structures include straight drop spillways, box inlet drop spillways, drop boxes, concrete chute spillways and drop inlet spillways

Grassed Waterway - A natural or constructed channel that is shaped or graded to required dimensions and established in permanent or continuous vegetation for the stable conveyance of run-off. The purpose of this structure is to convey run-off from terraces, diversions or other water concentrations without causing erosion of flooding and to improve water quality. Where existing vegetative cover is inadequate to protect the natural waterways or swales from erosion resulting from additional water volumes and velocities, the channels can be improved by grading, reshaping, fertilizing and revegetating. Man-made outlets should be designed to provide adequate capacities. They should be proportioned to carry run-off at safe velocities.

Heavy Use Area Protection - Protecting against erosion on heavily used areas by establishing vegetative cover, by surfacing with suitable materials, or by installing needed structures.

Lined Waterway or Outlet - A waterway or outlet with an erosion-resistant lining of concrete, stone or other permanent material. The lined section extends up the side slopes to the designed depth of flow.

The earth above the permanent lining may be vegetated or otherwise protected. It's purpose is to provide for the safe disposal of run-off from other concentrations of flow without damage by erosion or flooding, where unlined or grassed waterways would be inadequate

Mulching - Applying plant residues or other suitable materials, not produced on the site, to the surface of the soil. This application will help conserve moisture; prevent surface compaction or crusting; reduce run-off and erosion; control weeds; and help establish plant cover.

Pasture and Hayland Management - Proper treatment and use of pastureland or hayland. By use of a proper management system, one can prolong the life of desirable forage species; maintain or improve the quality and quantity of forage; and protect the soil and reduce water loss.

Pasture and Hayland Planting - Establishing and reestablishing long-term stands of adapted species of perennial, biennial or reseeding forage plants. This is done to reduce erosion, to produce high-quality forage and to adjust land use.

Roof Run-off Management - A facility for collecting, controlling and disposing of run-off water from roofs. This helps to prevent run-off water from flowing across concentrated waste areas, barnyards, roads and alleys, and to reduce pollution and erosion, improve water quality, prevent flooding, improve drainage, and protect the environment.

Run-off Management System - A system for controlling excess run-off caused by construction operations at development sites, changes in land use, or other land disturbances. This is done to minimize such undesirable effects as flooding, erosion, sedimentation and to maintain or improve water quality.

Sediment Basin - A basin constructed to collect and store sediment and debris. This practice applies where conditions preclude the installation of erosion control measures to keep soil in place or where a sediment basin offers the most practical solution to containing erosion.

Stripcropping

Contour - Growing crops in a systematic arrangement of strips or bands on the contour to reduce water erosion. The crops are arranged so that a strip of grass or close-growing crop is alternated with a strip of clean-tilled crop or fallow or a strip of grass is alternated with a close-growing crop. This done to reduce sheet and rill erosion, manage run-off to increase plant available moisture and improve surface water quality by reducing siltation.

Field - Growing crops in a systematic arrangement of strips or bands across the general slope (not on the contour) to reduce water erosion. The crops are arranged so that a strip of grass or a close-growing crop is alternated with a clean-tilled crop or fallow. This practice will reduce sheet and rill erosion and improve surface water quality by reducing siltation.

Terrace - An earth embankment or a ridge and channel constructed across the slope at a suitable spacing and with an acceptable grade. They are built to reduce erosion, reduce sediment content in run-off water, intercept and conduct surface run-off of a non-erosive velocity to a stable outlet, prevent gully development, improve farmability and reduce flooding, and increase soil moisture. For more complete information concerning design, construction, and maintenance of the preceding best management practices, refer to the current edition of the *Pennsylvania Soil and Water Conservation Technical Guide* prepared by the Natural Resources Conservation Service.

Definitions

Accelerated Erosion The removal of the surface of the land through the combined action of man's activities and the natural processes, at a rate greater than would occur because of the natural process alone.

Agricultural operations - The management and use of farming resources for the production of crops, livestock, or poultry

Best Management Practice (BMP) - Management activities, facilities, measures, or procedures used to minimize pollution to waters of the Commonwealth.

Commission - The State Conservation Commission established by the Act of May 15, 1945 (P.L. 547, No. 217), 3 P.S. §§849-864, known as the Conservation District Law, and as amended by Act 18 of 1995.

Concentrated water flow areas - Those natural or manmade areas where storm water run-off is channeled and conveyed directly to a surface body of water. Concentrated water flow areas include, but are not limited to, ditches, waterways, gullies, and swales.

Conservation Plan A plan that identifies Best Managaement Practices for agricultural plowing and tilling. A conservation plan which includes location, land use, tillage systems, and conservation treatment measures and schedules is considered to be an erosion and sediment pollution control plan.

Erosion - The natural process by which the surface of the land is worn away by the action of water, wind, or chemical action.

Farm - The land area devoted to an agricultural use, excluding greenhouses, mushroom production, or confined livestock or poultry operations

Implemented (also implementation) - Adequately applied or having a written schedule for implementation.

Natural Resources Conservation Service (NRCS) - The Natural Resources Conservation Service of the United States Department of Agriculture, formerly know as the Soil Conservation Service.

Pennsylvania Agronomy Guide - The quick reference book published by the Cooperative Extension as a practical guide to grain and forage production, soil fertility management, pest management, and erosion control, with special reference to Pennsylvania conditions.

Pennsylvania Erosion and Sediment Pollution Control Programme Manual - A primary reference document published by the Pennsylvania Department of

Environmental Resources, Bureau of Land and Water Conservation, which is used by technically trained persons to prepare erosion and sedimentation control plans for earth moving activities.

Pennsylvania Soil and Water Conservation Technical Guide - A primary reference document published by the United States Department of Agriculture's NRCS, which is used by technically trained persons to plan and apply BMPs.

Plowing and tilling - The disturbance of the land surface for an agricultural activity

Sediment - Soils or other surface materials transported by surface water as a product of erosion.

Sedimentation - The process by which sediment is deposited on stream bottoms.

Stormwater - drainage run-off from the surface of the land resulting from rain, or snow or ice melt.

Tilled area - Farm fields used for crop production that have less than 30 per cent crop residue cover during times of crop establishment, and less than 50 per cent cover during other times of the year. A cover crop may be used instead of crop residue.

Bibliography

A.L. Bhatia: *Handbook of Microbiology*, Pointer Publication, New Delhi, 2005.

AVERY, B.W., BASCOMB, C.L.: *Soil Survey Laboratory Methods*, Soil Survey Technical Monograph No.6, pp83. Harpenden, UK, 1982.

AVERY, B.W.,: *Soil classification for England and Wales*, Soil Survey Technical Monograph No.14, Harpenden, UK, 1980.

B.C. Mazumdar: *Orchard Irrigation and Soil Management Practices*, Daya Publications, Delhi, 2004.

B.P. Sawant, H.W. Awari, A.M. Kamble, D.D. Tekale: *Fundamentals of Soil and Water Conservation Engineering*, Current Publications, Agra, 2010.

Benu Singh: *Soil Management*, Vista International Publication, Delhi, 2008.

C.R. Sudharmai Devi: *Analytical Procedures in Soil Science and Agricultural Chemistry*, Agrotech Publisher, Delhi, 2004.

Charles Ernest Millar.: *Soil Fertility*, Biotech Books, Jaipur, 2004.

Charlotte Jhonson.: *Biology of Soil Science*, Oxford Book Company, Delhi, 2009.

D.A. Dodia, I S Patel and G M Patel.: *Botanical Pesticides for Pest Management*, Scientific Publication, Delhi, 2010.

Dewasish Choudhary: *Basics of Agricultural Chemistry*, Anmol Publication, Delhi, 2009.

H V Nanjappa, B K Ramachandrappa: *Manual on Practical Agricultural Meteorology*, Agrobios Publisher, Delhi, 2007.

K. Kolay: *Soil Fertility*, Atlantic Publications, Delhi, 2010.

K. Kumutha, K.G. Anitha, K. Saranya: *Agricultural and Industrial Microbiology*, Satish Serial Publishing House, Delhi, 2012.

L K Tripathi: *Water and Soil Management*, ABD Publication, Delhi, 2005.

M L Srivastava: *Soil Science*, Shree Publication, Delhi, 2007.

Malcolm D. Bolton: *A Guide to Soil Mechanics*, Universities Press, 2003.

Mangat Rai: *Soil Science*, Anmol Publication, Delhi, 2008.

N.K. Prasad: *Soil Fertility and Plant Nutrition*, IBDC Publishers, Oxford University Press, Jaipur, 2005.

Naren Kr. Dutta: *Principles and Practice of Soil Science*, Kanishka Publication, Delhi, 2005.

Parmeshwar Singh.: *Encyclopaedia of Agricultural Pests and Their Management*, Anmol Publication, Delhi, 2010.

Pravin Chandra Trivedi.: *Plant Pest Management, Aavishkar Publication*, Delhi, 2002.

R.J. Haynes.: *Mineral Nitrogen in the Plant: Soil System*, Academic Press is an imprint of Elsevier, 2013.

R.K. Kaleeswari, R. Rajeswari, J. Prabhaharan: *A Handbook of Soil Fertility*, Satish Serial Publishing House, Delhi, 2013.

R.K. Mehra: *Textbook of Soil Science*, Indian Council of AgriculturalResearch, 2006.

Rami Horowitz; Isaac Ishaaya.: *Insect Pest Management:Field and Protected Crops*, Springer, London, 2004.

Shivanand Tolanur.: *Fundamentals of Soil Science*, International Book Distributing Co, Delhi, 2004.

Shivan, Tolanur.: *Soil Fertility, Fertilizer and Integrated Nutrient Management*, International Publication, Delhi, 2006.

Shubhrata R. Mishra.: *Plant Protection and Pest Management*, Discovery Publishing House, Delhi, 2005.

T.V.S. Prasad.: *Soil Chemistry : Nutrient and Water Management in Agricultural Soils*, Dominant Publication, Delhi, 2009.

Tapan Adhikari, S. Kundu, A. Subba Rao.: *Nanotechnology in Soil Science and Plant Nutrition*, New India Publishing Agency, Delhi, 2013.

Index

I

L

M

P

R

S

T

U

V

W